ESCALATION AND INTRAWAR DETERRENCE DURING LIMITED WARS IN THE MIDDLE EAST

W. Andrew Terrill

September 2009

*****

I would like to thank Mary J. Pelusi, Lieutenant Colonel John A. Mowchan of the U.S. Army War College Center for Strategic Leadership, Dr. Steven Metz of the U.S. Army War College Strategic Studies Institute, Major David M. Burke, U.S. Air Force, and Sarah E. Womer for useful and insightful comments on earlier drafts of this work. My special thanks also go to Dr. Norman Cigar for insightful comments about my research as I went forward with this monograph. I am also grateful to Dr. Cigar for all that I learned from him when we served together at the Pentagon together during Operations DESERT SHIELD and DESERT STORM (myself as a mobilized reservist). All mistakes in this work of fact, omission, interpretation, and speculation are, nevertheless, entirely my own.

*****

Comments pertaining to this report are invited and should be forwarded to: Director, Strategic Studies Institute, U.S. Army War College, 122 Forbes Ave, Carlisle, PA 17013-5244.

*****

All Strategic Studies Institute (SSI) publications are available on the SSI homepage for electronic dissemination. Hard copies of this report also may be ordered from our homepage. SSI's homepage address is: *www.StrategicStudiesInstitute.army.mil.*

*****

The Strategic Studies Institute publishes a monthly e-mail newsletter to update the national security community on the research of our analysts, recent and forthcoming publications, and upcoming conferences sponsored by the Institute. Each newsletter also provides a strategic commentary by one of our research analysts. If you are interested in receiving this newsletter, please subscribe on our homepage at *www.StrategicStudiesInstitute.army.mil/newsletter/.*

ISBN 1-58487-406-6

CONTENTS

FOREWORD

The number of declared nuclear powers has expanded significantly in the last 20 years to include Pakistan, India, and North Korea. Additionally, other powers such as Iran are almost certainly striving for a nuclear weapons capability while a number of countries in the developing world possess or seek biological and chemical weapons. In this milieu, a central purpose of this monograph by W. Andrew Terrill is to reexamine two earlier conflicts for insights that may be relevant for ongoing dangers during limited wars involving nations possessing chemical or biological weapons or emerging nuclear arsenals. Decision-makers from the United States and other countries may have to consider the circumstances under which a smaller and weaker enemy will use nuclear weapons or other mass destruction weapons. Some of Dr. Terrill's observations may be particularly useful for policymakers dealing with future crises involving developing nations that possess weapons of mass destruction (WMD). Although it is possible that the United States could be a party to such a conflict, any crisis involving nuclear weapons states is expected to be of inherent concern to Washington, even if it is not a combatant.

Dr. Terrill has examined two important Middle Eastern wars. These conflicts are the 1973 Arab-Israeli War and the 1991 Gulf War. This monograph may be particularly valuable in providing readers, including senior military and political leaders, with a discussion of the implications of these historical case studies in which WMD-armed nations may have seriously considered their use but ultimately did not resort to them. Both of these wars were fought at the conventional level, although the prospect of Israel using nuclear weapons (1973), Egypt using biological weapons (1973), or Iraq using chemical and biological weapons (1991) were of serious concern at various points during the fighting. The prospect of a U.S. war with WMD-armed opponents (such as occurred in 1991) raises the question of how escalation can be controlled in such circumstances and what are the most likely ways that intrawar deterrence can break down. This monograph will consider why efforts at escalation control and intrawar deterrence were successful in the two case studies and assess the points at which these efforts were under the most intensive stress that might have caused them to fail. Dr. Terrill notes that intrawar deterrence is always difficult and usually based on a variety of factors that no combatant can control in all circumstances of an ongoing conflict.

The Strategic Studies Institute is pleased to offer this monograph as a contribution to the national security debate on this important subject as our nation continues to grapple with a variety of problems associated with the proliferation of nuclear, biological, and chemical weapons. This analysis should be especially useful to U.S. strategic leaders and intelligence professionals as they seek to address the complicated interplay of factors related to regional security issues and the support of local allies. This work may also benefit those seeking greater understanding of long range issues of Middle Eastern and global security. We hope this work will be of benefit to officers of all services as well as other U.S. Government officials involved in military planning, and that it may cause them to reconsider some of the instances where intrawar deterrence seemed to work well but may have done so by a much closer margin than future planners can comfortably accept. In this regard, Dr. Terrill's work is important to understanding the lessons of these conflicts which might otherwise be forgotten or oversimplified. Additionally, an understanding of the issues involved with these earlier case studies may be useful in future circumstances where the United States may seek to deter wartime WMD use by potential adversaries such as Iran or North Korea. The two case studies may also point out the inherent difficulties in doing so and the need to enter into conflict with these states only if one is prepared to accept the strong possibility that any efforts to control escalation have a good chance of breaking down. This understanding is particularly important in a wartime environment in which all parties should rationally have an interest in controlling escalation, but may have trouble doing so due to both systemic and wartime misperceptions and mistakes that distort communications between adversaries and may cause fundamental misunderstandings about the nature of the conflict in which these states may find themselves embroiled.

DOUGLAS C. LOVELACE, JR.
Director
Strategic Studies Institute

ABOUT THE AUTHOR

W. ANDREW TERRILL joined the Strategic Studies Institute (SSI) in October 2001, and is the General Douglas MacArthur Professor of National Security Affairs. Prior to his appointment, he served as a Middle East nonproliferation analyst for the International Assessments Division of the Lawrence Livermore National Laboratory (LLNL). In 1998-99, Dr. Terrill also served as a Visiting Professor at the U.S. Air War College on assignment from LLNL. He is a former faculty member at Old Dominion University in Norfolk, Virginia, and has taught adjunct at a variety of other colleges and universities. He is a retired U.S. Army Reserve lieutenant colonel and Foreign Area Officer (Middle East). Dr. Terrill has published in numerous academic journals on topics including nuclear proliferation, the Iran-Iraq War, Operation DESERT STORM, Middle Eastern chemical weapons, ballistic missile proliferation, terrorism, and commando operations. Since 1994, at U.S. State Department invitation, Dr. Terrill has participated in the Middle Eastern Arms Control and Regional Security (ACRS) Track 2 talks, which are part of the Middle East Peace Process. He also served as a member of the military and security working group of the Baker/Hamilton Iraq Study Group throughout its existence in 2006. Dr. Terrill holds a B.A. from California State Polytechnic University and an M.A. from the University of California, Riverside, both in Political Science. He also holds a Ph.D. in International Relations from Claremont Graduate University, Claremont, California.

SUMMARY

This monograph analyzes military escalation and intrawar deterrence by examining two key wars where these concepts became especially relevant—the 1973 Arab-Israeli War and the 1991 Gulf War against Iraq. Intrawar deterrence is defined as the effort to control substantial military escalation during an ongoing war through the threat of large-scale and usually nuclear retaliation should the adversary escalate a conflict beyond a particularly important threshold. The deep contrasts between the 1973 and 1991 dangers of escalation underscore the range of problems that can occur in these types of circumstances.

In the first case, this monograph relies upon an extensive body of openly available scholarship and investigative reporting on the 1973 War to discuss the potential for Israeli nuclear weapons use during that conflict. Although Israel is not a fully declared nuclear power, virtually all serious academic analysis both in and outside of that country assumes that there has been a strong Israeli nuclear weapons program for decades. Most major studies of the 1973 war suggest that Israel had or probably had some sort of nuclear option that it could have gone forward with in the event of an existential threat. Broad "hints" by the Israeli leadership, as well as their ongoing spending on nuclear research and nuclear-capable missile delivery systems, tend to support this. The work has proceeded on the assumption that the vast majority of scholarship about Israeli possession of a nuclear option during this conflict is correct, and that strong evidence included in this scholarship (which will be recounted here) suggests that Israel probably had a nuclear weapons option in 1973. In the very unlikely case that it did not, the Israelis probably had a different weapons of mass destruction (WMD) option that it could have used in conjunction with systems such as the *Jericho I* missile.

This work asserts that the Egyptians and the Syrians attacked Israel in 1973 with limited goals that included the capture of important territorial objectives but did not include the destruction of the Israeli state. Waging war into Israel itself beyond the range of their integrated air defense systems was beyond the capabilities of the Arab militaries, and they knew it. The Arab leadership appeared to believe that this situation should have been obvious to the Israeli leadership, but it was not. Although some military professionals such as then-Major General Ariel Sharon immediately understood the situation, others such as Defense Minister Moshe Dayan feared an existential threat. The sudden onset of a new war that began with a series of Arab battlefield victories deeply disoriented some Israeli leaders and appears to have pushed some into serious consideration of a nuclear solution. This outcome appears to have been avoided by the ability of Israeli leaders to discuss the threat in an open, professional, and democratic fashion which in this case allowed the most reasonable voices to come to the fore. The decisive Israeli battlefield victory of October 14 eliminated the need for Israel to consider nuclear weapons use, although the Egyptians then faced defeat themselves and signaled that they also had serious options for escalating the war.

The case of Iraq in 1991 is also in need of some further examination primarily because the war itself was such a one-sided military victory, and the United States seemed almost effortlessly to deter Saddam Hussein from the use of his chemical and biological warfare options. The negative aspect of this very positive outcome is that there is some need to prevent this case from becoming too dramatic of a "false positive" of the ease in which intrawar deterrence can be implemented. Saddam Hussein throughout the crisis strongly believed that he would be able to fight the coalition troops to a standstill in conventional combat, and that this outcome would allow his regime to remain in place. With this deluded, but very real, conventional strategy for victory, he was reluctant to escalate beyond the conventional level where he expected Iraq to do very well. While Saddam Hussein feared escalation to nuclear weapons use by the United States, his tendency to be deterred was bolstered by his perceived conventional options. He might have become more reckless if he was fully cognizant of the conventional strength of the U.S. military which he dismissed as having less fighting spirit than the Iranian troops that Iraq had previously defeated.

A central conclusion of this monograph is that intrawar deterrence is an inherently fragile concept, and that the nonuse of WMD in both wars was a result of factors that may or may not be repeated in future conflicts. Additionally, the tactics for intrawar deterrence will require constant adjustment as war is waged and develops in unexpected ways. Signaling and political communication is inherently difficult in such crisis and few unequivocal statements are taken at face value. U.S. planners must never become too comfortable with the elegance of any plan involving intrawar deterrence, and the U.S. leadership must be prepared to accept the possibility that there are always a number of ways such strategies can break down.

ESCALATION AND INTRAWAR DETERRENCE DURING LIMITED WARS IN THE MIDDLE EAST

INTRODUCTION

The October 1973 Arab-Israeli War (known in Israel as the Yom Kippur War and in the Arab World as the Ramadan War) and the 1991 U.S.-led Gulf War against Iraq (Operation DESERT STORM) are two very different types of regional conflict which each provide important examples of the dangers of military escalation resulting from mistakes and miscommunications inherent in rapid wartime decision-making during times of extreme stress. In both cases, one side of the conflict was determined to fight a conventional war with limited but important strategic objectives, while also deterring their opponent from escalating to unconventional weapons. The Egyptian/Syrian coalition fighting Israel in 1973 sought to wage conventional warfare against the Israelis to recapture some of the land that Israel had seized in the earlier 1967 war, but they also sought to avoid provoking Israel into use of its suspected nuclear weapons arsenal or triggering what they felt was the danger of U.S. military intervention to help the Israelis.[1] Of equal complexity, the U.S.-led coalition opposing Iraqi president Saddam Hussein in 1991 sought to expel Iraqi troops from Kuwait, and by doing so create a climate favorable for an Iraqi military coup, without provoking Saddam into using his chemical or biological weapons on either coalition troops and countries or on Israel. Both the Arab states in 1973 and the United States in 1991 sought to wage a war with limits that were favorable to them and communicate to their adversary that it was in everyone's interest to respect such limits. This task appears to be staggeringly complex in an environment distorted by rapidly changing events and the often limited ability for each side to understand the actions and motivations of adversaries.

The 1973 war is a particularly valuable example of the dangers of rapid military escalation that, according to many plausible sources, involved serious Israeli consideration of nuclear weapons use. The apparent reason for the danger of nuclear weapons use involved serious Israeli military setbacks at the outset of the war, which produced an environment where much of Israel's military doctrine was proven inadequate or even wrong, and the path to conventional military victory was no longer clear. Adding to Israeli concerns, the Arab armed services were then undertaking a variety of military activities that had previously been considered beyond their capabilities, and the Israelis consequently had ample reasons to doubt previous assessments on these matters. Moreover, some of the Israeli leadership initially assumed that they were confronted with an existential threat and that their country faced a serious chance of being overrun by Arab forces. Their special concern about the destruction of the state was grounded in a deep psychological sense of tragedy over the previous ghastly chapters in the history of the Jewish people and especially the Nazi Holocaust.[2] In retrospect, it is clear that Egyptian President Anwar Sadat and other Arab leaders viewed overrunning Israel as completely beyond their capabilities, but the Israelis may have misinterpreted the limitations of Arab military forces in the crisis of the moment. The Arab offensives into Sinai and the Golan Heights were further initiated with a successful surprise attack, giving the Israelis less time to prepare themselves for a military and political response to vastly improved Arab militaries using new strategy and tactics. The Israeli military nevertheless did turn the situation around to the point that President Sadat tried to protect Cairo's wartime gains by his own threats of escalation which were vague but potentially quite serious.

The 1991 Gulf War was a notably different type of conflict which has distinct but equally valuable lessons regarding military escalation and especially wartime deterrence of chemical and biological weapons use. Military escalation was a serious possibility in this instance for significantly different reasons. Saddam Hussein's 1990 invasion and seizure of Kuwait produced an immediate and escalating Iraqi political confrontation with the United States and many of its allies. In this case, communications between the hostile parties were more straightforward than in 1973, and one serious ministerial level diplomatic encounter (between U.S. Secretary of State James Baker and Iraqi Foreign Minister Tareq Aziz) occurred during the period of crisis shortly before the war itself. While the 1973 Egyptian/Syrian attack was a surprise attack with a complicated deception plan to obscure Arab intensions, Iraq anticipated the 1991 war and had been warned about the consequences of military escalation with chemical or biological weapons after war had broken out. Nevertheless, serious problems existed in communicating with Iraqi president Saddam Hussein because of the nature of the regime rather than the sudden onset of an unexpected war. While Israeli national culture (often including military culture) permits individuals to speak their minds, to do so in Iraq was usually unwise and in certain cases could be fatal. Saddam Hussein could and frequently did misunderstand Iraq's strategic situation because of his own poorly formulated view of key international and military factors and events. Usually no one around him dared to contradict or correct the Iraqi dictator, thereby allowing his misperceptions to go largely unchallenged.

THE CONCEPT OF INTRAWAR DETERRENCE

The concept of intrawar deterrence involves a process of explicit or tacit bargaining within an on-going war that still has key limits or thresholds that have not been crossed. According to pioneering deterrence theorist Thomas Schelling, two fundamental issues are bargained over in limited war. These are the outcome of the war and the mode of conducting the war.[3] In this regard, most of the analysis of intrawar deterrence has been developed within a Cold War context. If, for example, the United States and the Soviet Union had fought a limited war in Europe, how could this be structured so that it did not escalate into a general war involving nuclear strikes against European population centers? If a war went nuclear in Europe, what would be the likelihood of that war escalating to involve an exchange of strategic weapons by the United States and the Soviet Union against their respective national homelands? Such questions are important within the literature of Cold War deterrence and escalation, but there is an important distinction between this form of deterrence and that often found in regional conflicts. U.S.-Soviet deterrence theory often envisions adversaries of roughly equal strength. Adversaries in regional conflicts may have some deterrent capacity, but it is often well below that of their adversaries. A regional power attempting to deter an attack by the United States (such as Iraq in 1991) is, of course, dramatically weaker than its adversary.

Deterrence with weapons of mass destruction

(WMD) is usually described as an effort by one country to threaten another with dire consequences if the adversary embarks on a particular course of action. In this regard, WMD are generally described as including nuclear weapons, biological weapons, and some of the more effective forms of chemical weapons. In any deterrent relationship, a deterring nation must often convince its adversary that it has both the capability and the will to utilize strategic weapons if certain thresholds are crossed. These thresholds are sometimes described as "red lines." The adversary is expected to refrain from certain actions because a rational cost-benefit analysis is formally or informally performed and the relevant leadership correspondingly understands that some actions will lead to impermissible consequences. Therefore restraint must be employed. In the Cold War context, the weapons used for deterrence were almost always nuclear weapons (although some international relations theorists have considered the question of conventional deterrence).[4] In regional conflicts, nuclear weapons are often not available to all combatants. This situation has sometimes led to efforts to impose strategic deterrence with other unconventional weapons including biological weapons and sometimes chemical weapons.

Rationality is often considered a prerequisite to effective deterrence. Under this interpretation, an adversary must be able to perform at least a crude cost-benefit analysis and then be guided into making the most clearly reasonable choice given the available options. Assuming some degree of rationality is usually inherent in applying deterrence as a tool for preventing enemy actions. Nevertheless, not all deterrence theorists agree that rationality is required for effective deterrence. Some scholars state that fear can be the dominant component of the deterrence equation and that fear is not rational. Deterrence theorist Patrick Morgan underscores this point by asserting that small children, animals, and mentally ill people can all be deterred from disapproved courses of action, although this process occurs without a comprehensive cost benefit analysis.[5] If an opponent has only to have the "rationality" level of a dog, it is possible that in some circumstances the need for rationality is overemphasized. Unfortunately, there is also a catch in this logic since fear can cause people to behave in unpredictable and counterproductive ways. While a frightened leader of limited rationality may mimic a reasonable approach to strategic problems out of fear, that same leader may undertake counterproductive responses to a strategic crisis out of that same fear. If fear rises to a level near panic, the tendency to make bad decisions may rise accordingly.

Intrawar deterrence is probably the most difficult form of deterrence to implement since a state pursuing such a policy is waging war against another nation while seeking to prevent its opponent from responding with all of the weapons that it possesses. Such a task is, to say the least, challenging since both sides usually seek to use as much of their capabilities as possible to optimize their chances of victory. Often wars involving intrawar deterrence have limited objectives such as in the two examples that this study will discuss. In such circumstances, rationality, to the extent it is present, may still be overwhelmed by stress, panic, or a retreat into core religious or ideological values that may be inappropriate to the situation. The key to addressing this problem may not be to assume rationality. It may be to do as much as possible to present threats in a way that channel the enemy's fear into specific directions. This may be increasingly possible under the circumstances of asymmetric deterrence, where a weaker power attempts to deter a much more powerful adversary with a limited arsenal and may be deeply uncertain if this is achievable.

There are also some special problems with Israel's approach to deterrence that are relevant to this analysis. The Israeli government has only obliquely admitted that it possesses a nuclear weapons capability and does not permit its citizens to publish information about the technical capabilities of these weapons.[6] This policy is called "opacity" or "nuclear ambiguity."[7] While Israeli military censors do not allow Israeli citizens to discuss the numbers, characteristics, or deployment of Israeli nuclear weapons, they do accept academic discussions about the nature of deterrence options which such weapons may provide to Israel. It should also be noted that a number of Israeli scholars have provided thoughtful and valuable analysis of their own country's deterrent efforts in more general and theoretical terms.[8] This analysis has dealt extensively with both the 1973 war and Israel's role in the 1991 Gulf War, but the details of how an Israeli nuclear response might have taken place in these conflicts have been left vague and hypothetical in many instances. Nevertheless, a precise understanding of Israeli nuclear capabilities in 1973 or the nature of the Iraqi biological/chemical warfare threat is probably less important for this analysis than a consideration of how these capabilities were perceived by the relevant opposing leaders at the time and how these perceptions influenced risk-taking and other wartime behaviors.

THE 1973 ARAB-ISRAELI WAR

Background.

The conditions for the October 1973 War were set 6 years earlier with the decisive Israeli victory in the June 1967 Six Day War. During the 1967 conflict, Israel easily defeated the combined military forces of Egypt, Syria, and Jordan and in the process captured significant amounts of territory from all three countries.[9] A large Iraqi army expeditionary force was also destroyed as an effective fighting force by the Israeli Air Force (IAF) before it was able to enter ground combat.[10] The June 1967 War created a new sense of optimism in Israel and a strategic outlook that virtually defined decisive victory over Arab forces as a part of the Israeli national identity. Israel's 2-year economic recession ended almost immediately in the aftermath of the war.[11] Foreign capital and new immigrants started to flow much more freely into Israel after the victory. Israel's future existence no longer seemed to be uncertain, and Israeli Defense Minister Moshe Dayan became an internationally known symbol of Israeli military prowess. In Egypt, by contrast, a number of senior Egyptian officers found themselves on trial for criminal negligence because of the magnitude of the defeat. The Egyptians had led the Arab coalition against Israel, and the Egyptian military was profoundly humiliated by the scale of their defeat.[12]

In the aftermath of the June 1967 conflict, few Israeli leaders felt a serious need to negotiate a political compromise with the defeated Arabs. There seemed to be little to worry about from a future attack by the Arab militaries, and thus no serious national security penalty for ignoring Arab demands. In any event, such demands were often excessive and usually stressed the need for Israel to the return of all occupied territories without any comprehensive diplomatic, economic, or political concessions being granted to Israel in exchange. Such terms were unacceptable, and many Israeli leaders believed that time was on their side because of what they saw as Arab obstinacy. The longer the Arabs waited to negotiate seriously, the less there would be to negotiate over as Israel built Jewish settlements in the territories, including the Sinai, in what was widely regarded as an attempt to extend Israeli sovereignty to the captured territory. This approach assumed continuing Israeli military domination of the region, which was often viewed as beyond serious challenge. Israeli policymakers and especially the military leadership also felt little need to question the effectiveness of the strategy and tactics of the June 1967 War or refine military doctrine extensively on the basis of potential improvement in Arab armies. While defeat can often be a powerful instructor, this is seldom the case with victory.

As a result of the 1967 victory, Defense Minister Dayan no longer believed that Arab adversaries were capable of seriously threatening Israel or dominating the escalation process in conventional war. Israel's armor heavy formations which minimized the use of infantry support were hailed as rewriting previous lessons of modern maneuver warfare, while airpower was seen as something of a panacea which negated the need for a variety of ground capabilities including substantial artillery support. Moreover, the limited Egyptian-Israeli military confrontation known as the 1969-70 "War of Attrition" did little to change Israeli minds about Arab capabilities. This conflict was characterized by both Egyptian and Israeli commando strikes across the Suez Canal as well as heavy Egyptian artillery bombardments of Israeli positions in an effort to pressure the Israeli commanders into withdrawing from western Sinai. The struggle ended following an Israeli decision to punish Egypt with "deep penetration raids" by Israeli aircraft striking at economic infrastructure targets in Egypt. These raids were widely credited in Israel as forcing Egypt to accept a cease-fire in August 1970.[13]

The aftermath of the June 1967 War also led to a more rigid approach to Israeli strategic thinking which was increasingly dominated by an outlook known as "the concept." In its most straightforward form, the concept stated that Syria would not attack Israel unless it did so in collaboration with Egypt, while Egypt would not attack Israel until it had achieved air parity with Israel.[14] Conveniently, Egyptian air parity with Israel seemed elusive and perhaps unobtainable. The IAF was one of the finest air forces in the world, and was equipped with some of the best available U.S. and French aircraft. Additionally, Israeli pilots routinely outperformed Arab pilots and had on at least one occasion even shot down Soviet pilots flying in Egypt's Soviet-made aircraft during the War of Attrition. This success was achieved without losses on the Israeli side.[15] In general, however, the Israelis made strenuous efforts to avoid hostile aircraft if the pilots were engaging in radio traffic in Russian because they feared that humiliating the Soviets could cause Moscow to provide additional support to Arab states. In this one instance, that rule was broached long enough for the Egyptians to realize that they were receiving air combat instruction from a country whose pilots had not proven up to dealing with the Israelis themselves.[16] Israel's air superiority seemed increasingly beyond challenge.

Israeli military doctrine maintained an overwhelming faith in the value of Israeli air and armored superiority that seemed immune to the concept of tactical innovation by the other side in ways which might allow them to offset their shortcomings in these forms of warfare. Some Israeli officers (and particularly junior officers) did question core strategic beliefs in light of new Arab acquisitions of significant numbers of modern anti-aircraft and anti-tank weapons, but they were a distinct minority that often felt the disapproval of their superiors.[17] In one of the most important books on the 1973 War, author Abraham Rabinovich stated that senior intelligence officials, "Explain[ed] away every piece of information that conflicted with their thesis, [while] they embraced any wisp that seemed to confirm it."[18] The Arab states, however, understood that they would almost certainly not be able to overcome Israeli air superiority in the foreseeable future by building up their air forces. Neither the Egyptians nor Syrians had been able to match Israeli standards in the period prior to the 1967 War. In that war, Arab airpower received a staggering blow when 286 Egyptian military aircraft were destroyed on the ground in the first few hours of the war with many of the remaining aircraft destroyed later in the conflict.[19] The Jordanian and Syrian Air Forces were also almost completely wiped out. A large number of Arab pilots were also killed in the fighting, depriving the Arab states of an important experience base which had to be rebuilt.

The Israeli devotion to a rigid and potentially flawed strategic doctrine was a military weakness that was targeted for exploitation by opposing forces. On the eve of war, the Egyptians and Syrians were aware of this situation and correspondingly implemented a comprehensive deception plan tailored to play to

Israeli biases about the shortcomings of Arab armies. The massing of Egyptian forces along the Suez Canal just prior to the war was presented to the world as part of the "Tahrir 41" military maneuvers which were supposedly scheduled for October 1-7. The arrival of additional Egyptian troops along the Suez Canal was correspondingly expected, and Israel did not consider this provocative or dangerous. Additionally, a number of Egyptian soldiers were assigned to help deceive the Israelis by appearing lackadaisical, engaging in such activities as fishing, sunbathing, and sitting at their posts without helmets or appropriate military gear.[20] In the north, a Syrian decision to mass troops was widely viewed as posturing in the aftermath of a September 13 air incident over Lebanon in which the Syrian air force challenged an Israeli combat air patrol aircraft and then was decisively defeated, with 12 Syrian aircraft shot down while Israel only lost one plane.[21] Egypt and Syria were therefore able to mass their forces in a way that did not unduly alarm the Israelis.

In support of the deception plan, Arab radio traffic on both fronts was kept to a minimum and high ranking political and military leaders were seen keeping routine schedules. In what appears to be a special added touch, in late September a Syrian-supported terrorist organization took five Jewish hostages on their way by train to an Austrian transit center which helped to facilitate the movement of Russian and Eastern European Jews to Israel. When the Austrian government agreed to close the Schonau Castle transit point in exchange for the release of the hostages, the agreement produced a major international uproar that strongly distracted the attention of the Israeli leadership. Against this background, a number of strikingly clear intelligence indicators were ignored because the Israelis were not prepared to accept that they indicated a serious possibility of war. The worst example of an important intelligence indicator that was minimized by Israeli military intelligence was a large-scale evacuation of Soviet citizens from Syria on the eve of war. Arab deception and Israeli self-deception consequently led to a massive intelligence failure.

The Egyptian/Syrian Decision to Initiate War and the Efforts to Control Escalation.

The Egyptian decision to go to war with Israel came as a result of the Cairo leadership's belief that the status quo, and particularly Israel's occupation of Egypt's Sinai Peninsula, was intolerable. War against Israel was a difficult and dangerous option for Egypt. The chance of yet another humiliating defeat was always there, although the Egyptians hoped that they could minimize this possibility with methodical preparation and limited objectives. The Egyptian willingness to accept a limited war with carefully proscribed goals was linked to a number of factors, one of which was suspicion of the existence of Israeli nuclear weapons while others included the limitations of their logistical infrastructure and problems that could arise once Egyptian forces began operating beyond the range of the air defense systems located on the Suez Canal. A different kind of war with the idea of overrunning Israel would clearly raise the prospect of a nuclear strike if Israel actually had the suspected nuclear weapons at this time.[22] Even in the absence of nuclear weapons, Sadat may have believed that the United States would intervene with military force before it would allow Israel to be defeated and overrun.[23] The Egyptians used the codename "Operation Spark." The spark indicated the beginning of a process that began with a military option and ended with a political solution brokered by the superpowers. Three days into the war, Sadat secretly signaled Secretary of State Henry Kissinger that he wanted post-war negotiations leading to a permanent settlement. Kissinger is reported to have commented, "This is a statesman who understands that diplomacy is the other side of the battlefield."[24]

After the war, Cairo used the political process to restore the Sinai Peninsula to Egypt in an effort which culminated in the Egyptian-Israeli peace treaty of March 1979. Nevertheless, in October 1973, it was by no means certain that Sadat had decided that formal peace with Israel would be acceptable. He wanted Israel to withdraw from the Sinai and believed that a new relationship with the United States following a victorious war was the best way to achieve that goal. In 1973, he may have hoped that the scope of his victory would be so sweeping as to lead to a U.S.-brokered cease-fire, which returned the Sinai to Egypt but did not require Egypt to enter into a formal peace treaty with Israel. If this was the case, his strategic thinking involving Israel evolved quite dramatically in the post-war period. Another point that is frequently made when analyzing Sadat's mindset is that no Arab leader wanted to negotiate from weakness in which the dismal performance of 1967 had robbed them of both dignity and the military credibility that can be indispensible in pursuing successful negotiations with a tough adversary while maintaining at least some elite and public support for your approach.

In 1973, the Syrians were equally committed to recovering their territory lost in 1967, but they were particularly unwilling to pursue this option through direct negotiations with Israel at that time. The

Syrians, in contrast to the Egyptians, hoped to take back all of the territory that Israel had captured in 1967 by military means and had little interest in a follow-on diplomatic process.[25] Damascus was also unable to consider a military option unless a Syrian attack on Israeli positions in the Golan was part of a much larger war that included Egypt (as Israel's "concept" doctrine correctly pointed out). Before agreeing to a war with Israel, Syria also needed to ensure that Cairo was fully committed to a large-scale conflict that bogged down substantial numbers of Israeli military forces committed to the Egyptian front. These troops would have to be focused on fighting Egyptian forces all or at least most of the war if Syria was to have any chance of meeting its goals. Sadat's interest in gaining the Syrians as allies in the upcoming war was important if Egypt was to avoid the full fury of an Israeli response to its forces crossing the Canal. To gain Syrian support and involvement, he engaged in what one author has somewhat kindly referred to as a "deception campaign" against the Syrians.[26]

The Egyptian effort to bring Syria into the war included falsely briefing Damascus on Egypt's "Granite I" war plan as the basis for the attack. This deeply maximalist plan involved an initial thrust deep into Sinai to seize the strategic Mitla and Gidi passes and then press on to seize the entire Sinai Peninsula. Granite I had previously been disregarded as unrealistic by the Egyptian High Command, but it presented a level of Egyptian military commitment that was deeply reassuring to Damascus.[27] Syria agreed to be part of the war on the basis of the Granite I plan with the objective of seizing the Golan Heights. The Syrians had no operational plans to continue the assault down the Galilee after this goal was accomplished.[28] They also did not seem to have the logistical system to do this, even if the IAF could be neutralized first. Shortly after the war started, the Syrian leadership realized that the Egyptians had deceived them about the scope of the war, but there was little that they could do about it. After the war, Syrian President Hafez al-Assad told King Hussein of Jordan that he had been "exploited" by Sadat.[29]

Even with limited territorial objectives, the Arab states had to come to grips with the issue of Israeli nuclear weapons. In the past, various Arab leaders, including Sadat, had asserted that Israelis did not yet have nuclear weapons but sought to suggest that they did in order to intimidate the Arab states.[30] Such assertions may have been useful for public declarations to keep morale high, but they were difficult to accept as the basis for military planning. The Israelis had possessed a French-supplied nuclear reactor in the Negev desert at Dimona, which became operational in 1964. This reactor had the capacity to produce militarily significant amounts of weapons grade plutonium provided that the Israelis could reprocess the spent fuel. Avner Cohen, the leading Israeli scholar on the country's nuclear weapons program, estimates that Israel probably had the ability to use at least one, and probably two, nuclear weapons in the June 1967 War had it needed them.[31] This understanding seems to coincide with the U.S. view at the time. According to a variety of newly declassified documents, the U.S. Government was conducting foreign policy in the late 1960s in the belief that Israel either had nuclear weapons or could assemble them on short notice.[32] The overwhelming Israeli conventional victory in that conflict meant that the Israelis never had to consider nuclear weapons use since their victory was rapidly and easily obtained without such systems.

Israel had also refused to become a party to

the 1970 Nuclear Nonproliferation Treaty, and the Israeli leadership did not allow inspectors from the International Atomic Energy Agency (IAEA) to visit the Dimona reactor.[33] Many Israeli commentators also quite logically asserted that a small country like Israel could not indefinitely maintain conventional military superiority over all neighboring Arab states, which might become involved in a military confrontation. If any more evidence of Israel's interest in nuclear weapons was needed, it might be considered that Israel, despite a huge and public investment in nuclear technology, has never maintained or even seriously considered developing a civilian nuclear power program to generate electricity for its cities. Moreover, Egypt was especially vulnerable to a nuclear countervalue attack since the destruction of the Aswan dam with nuclear weapons could lead to massive numbers of casualties. Egypt's Aswan Dam is the largest rock filled dam in the world and is widely regarded as impossible to destroy without the use of nuclear weapons. If the Israelis felt that their country was about to be destroyed, Cairo would be ill advised to count on Israeli restraint.

Under the above circumstances, it seemed prudent for the Arab states to take the Israeli nuclear option seriously in their strategic planning. If there was even a limited chance of Israeli nuclear weapons use, the Egyptians and the Syrians would have to anticipate ways to compel Israel to fight at the conventional level through whatever means possible. The most logical way to do this would be to generate a serious threat of inflicting mass casualties should Israel choose to escalate the conflict to nuclear weapons use. As late as 1972, it was not immediately apparent how the Arabs would be able to present such a challenge, but it would have to include a threat of Egyptian unconventional weapons and the means to deliver these weapons. Turning first to potential delivery systems for unconventional weapons, the Egyptians had previously attempted to build their own large-scale ballistic missiles with the aid of German missile engineers, but this effort turned out to be a spectacular failure.[34] The huge *Kahir* surface-to-surface missile, while equipped with a destructive conventional warhead, had a range of only a few miles and was wildly inaccurate. Lieutenant General Saad al Shazly, Egypt's Army Chief of Staff in 1973, noted that politicians liked to boast about the missile and its smaller and equally ineffective counterpart, the *Zafir*, but these systems were essentially useless. Shazly characterized the lies and wasted millions of dollars surrounding the project as "shameful."[35]

Deterrence of Israeli escalation options through the use of long-range Arab bombers alone was also problematic because of Israeli air superiority and the limited prospect that Egypt's long-range bombers (which were older Soviet supplied systems) could penetrate Israeli air defenses through the use of their *Kelt* stand-off missiles. In a November 1972 discussion of the role of these missiles with the head of the American interests section in Cairo, Lieutenant General Shazly stated, "[Y]ou must understand that strikes against the interior of Egypt will now be met by strikes against the interior of Israel." These statements were made in the presence of President Sadat who quickly agreed with them.[36] Sadat, however, might have been somewhat concerned when his then air force commander, Lieutenant General Ali Baghdadi, informed him that the missiles were slow enough to be extremely vulnerable to a variety of Israeli countermeasures.[37] They were thus too unreliable to be the centerpiece of an intrawar deterrent strategy regardless of what kind of warheads were fitting on them. At best, the *Kelt* missiles seemed to be a system that had limited capabilities and could only be treated as a small part of an overall intrawar deterrence strategy.

The optimal form of delivery vehicle was a long-range missiles with a capability to strike into the Israeli heartland without the possibility of being intercepted by Israeli air defenses. The Soviets had been extremely reluctant to supply such weapons due to a fear that Western powers would hold them responsible for escalating the potential for war in the Middle East. Moscow's restraint was anathema to the Egyptains and eventually led President Sadat to expel the majority of Soviet advisors in July 1972.[38] In a letter to President Leonid Brezhnev explaining the decision, Sadat stated:

> In our repeated discussions, I mentioned that we needed deterrent weapons to make the enemy hesitate to strike deeply within our territory (as has been done in the past) knowing that we could, in turn, reach its own heartland. It was obvious then, and still is, that without these weapons we will not be able to act decisively and Israel will therefore find it unnecessary to change its stubborn position with regard to a solution.[39]

While Sadat never mentioned WMD in the exchange, he knew long-range missiles with conventional warheads were not much of a deterrent threat. The payload of one World War II B-17 bomber was significantly more than that of a *Scud*.

In the face of unrelenting Egyptian pressure, Moscow finally agreed to transfer two brigades of *Scud-B* missiles to Egypt, but these systems were only armed with conventional warheads.[40] This agreement appears to have been made in October 1972 during a visit by Egyptian Prime Minister Aziz Sidqi to Moscow, Russia.[41] The first *Scud* brigade arrived in Egypt on August 24, 1973.[42] These were the first such missiles that the Soviet Union had transferred to a non-Warsaw Pact country. The second brigade arrived before the war, but the exact date has not been established. On the eve of war, Egypt possessed at least 18 *Scud* missiles and nine launchers with Egyptian crews trained to operate them.[43] Syria had also requested *Scuds* from the Soviet Union but did not receive them until after the war. The Syrians did have short-range battlefield systems known as Free Rocket Over Ground 7s (FROG-7s), but these systems could not be used as a strategic threat due to their limited range and the conventional warheads supplied to the Syrians.[44]

The other aspect of a serious deterrent capability was to convince the Israelis that Egypt possessed a payload for the *Scuds* that would allow them to be utilized as strategic weapons able to inflict mass casualties. Prior to the war, the Egyptians attempted to indicate this by claiming a biological warfare (BW) capability. Biological weapons include systems that produce casualties through bacteria, viruses, or the toxins associated with living organisms.[45] In the early 1970s, Sadat announced that Egypt possessed a BW option which it would use against Israel in response to an Israeli BW attack, but the nature of the Egyptian biological agent that would be used in such circumstances was never clarified. In answer to a journalist's question on this subject, the Egyptian president stated, "I believe that the density of the Israeli population confined in a small area would provide the opportunity to reply with [a biological weapon] if they should begin using it. Briefly, we have the instruments of biological warfare in the refrigerators, and we will not use them unless they begin to use them."[46] In the unlikely event that the Israelis missed the implications of Sadat's warning, which was published in Egypt's leading newspaper *al Ahram*, an Egyptian deputy prime minister reiterated the threat in April 1972.[47] The Cairo leadership may have expected that such claims would give the Israelis pause about using any WMD (including nuclear weapons) even if the Egyptian BW capability remained unclarified. The Arab leaders may have further believed that their special relationship with the Soviet Union would serve as a deterrent to Israeli nuclear weapons use in anything except an existential war. Neither one of these options appears concrete enough to deter the Israelis with absolute certainty. Threatening either option could potentially be considered to be a strategic bluff that the Egyptians might not have wanted to push too far by appearing to threaten the existence of the state.

In contrast to Egypt, the Syrian capacity to strike back at Israel with strategic weapons was so clearly nonexistent that Damascus did not even offer a pretense that such an option was available. In response to this weakness, Syria sometimes appeared more cautious than Egypt about avoiding ways of provoking Israel into escalatory acts, and there were, as noted, no Syrian operational plans for continuing the war into Israel. This strategy may have been a function of their limited logistics, or Damascus may have been more cautious because of its lack of an indigenous deterrent capacity. Syria, as noted, attempted to acquire *Scuds* from the Soviet Union prior to the 1973 War but was unable to do so until 1974. According to one source, the Syrian preference expressed to Egypt at the beginning of the war was that Israel should not be bombed within the 1967 borders.[48] In the event of Israeli escalation behavior, Damascus would have to depend upon Egypt or the Soviet Union to engage in a policy of extended deterrence, something that was not clear that either ally was prepared to do in any reliable way.

The Course of the Fighting and the Israeli Struggle to Define Military Options.

The 1973 Arab-Israeli War began at 1:55 p.m. local time on Saturday, October 6, 1973. Simultaneous attacks were launched by the Egyptians across the Suez Canal and the Syrians on the Golan Heights. The attack across the Suez Canal was complex and well-organized, with the first wave of Egyptian forces crossing in rubber boats and then bypassing the strongpoints along Israel's Bar-Lev Line to prepare for an armor-heavy Israeli counterattack. They planned to defend against such a counterattack with handheld anti-tank weapons. To enable the main body of troops to cross the waterway, Egyptian combat engineers used high velocity water pumps to clear pathways along the sand banks where bridges could be emplaced. The Israeli defensive plan designed for such a contingency (code named Dovecote) envisioned limited numbers of Israeli troops supported by tactical air strikes containing the Egyptians until significant numbers of reserve forces were mobilized and moved into the battle, thus allowing Israel to go on the offensive. Only limited numbers of troops were deployed in support of this plan. Major General Albert Mendler, the commander of the 252 Armored Division (the only major force in Sinai), believed that if war broke out, it would be a new war of attrition involving only small scale raids and artillery bombardments.[49]

The Syrian strategic objective was to seize the Golan

Heights from Israel within the first 24 hours of the attack. After that, they hoped to work with the Soviets to obtain a United Nations (UN) sponsored cease-fire that would allow them to keep the Heights before the Israelis had time to launch a successful counterattack.[50] The possibility of seizing the Golan early in the conflict seemed plausible. On the Golan front, Israeli resources were even more limited than in Sinai, with only 177 Israeli tanks and around 200 infantrymen facing a Syrian force of 1,400 tanks and 40,000 troops. According to one author, such a limited deployment represented total Israeli contempt for the Syrian military more than it represented Israeli manpower limitations.[51] While the Syrians were not able to seize the entire Golan Heights as planned, they made significant territorial gains in the initial battle, and Israeli forces defending the area became the top priority for reinforcement by mobilized Israeli reserve forces.

The ability of the Arab forces to launch an attack that the Israelis only detected shortly before the war began meant that Israel did not have time to mobilize and deploy substantial numbers of reservists to repulse the initial strikes. Some Israeli leaders had even opposed the total mobilization of reserve forces on the eve of battle on the assumption that the Egyptians would be defeated before the entire reserve force could be brought into action.[52] Israeli leaders were surprised that the Egyptians had been able to cross the Suez Canal in force, but few of them saw either front as unmanageable despite the early setbacks. Israeli Defense Minister Dayan had previously predicted that any Egyptian attack across the Suez Canal would be defeated within 24 hours.[53] The general assumption on the Golan Heights was that they were facing a 1-day event as well.[54] Some Israelis referred to the expected fight as the "7th day" of the Six Day War, which would be dominated by armored forces and airpower. This viewpoint was a serious misreading of the operational situation. The restricted Egyptian objectives did not require them to employ an air force that was able to equal the IAF. Rather, Egyptian military forces only had to be able to control the airspace over a limited amount of territory where they would be waging ground operations. This goal was achieved with an integrated air defense system including surface-to-air missiles (SAMs) and anti-aircraft artillery (AAA) guns. Additionally, the Egyptians did not base their ground tactics on the tank-to-tank battles, at which the Israelis excelled. Rather, they stressed use of infantry in an anti-tank role on the assumption that the then-small Israeli artillery forces would be unable to counter such tactics.

The Israelis met the Arab advances with insufficient force and inadequate military doctrine. The defensive Bar-Lev Line in the Sinai failed to contain the Egyptian advance because it was not designed to do so without the significant reinforcements. The Dovecote plan assumed the military leadership would have at least 48 hours notice of an imminent attack, during which time reinforcements could be deployed. Instead, the undermanned, unreinforced, and besieged forts became bait, luring units of Major General Mendler's Sinai tank division into costly and usually unsuccessful rescue efforts. By the 4th day of the war, 49 Israeli aircraft, one-eighth of their air force, had been shot down.[55] On the ground, Mandler's division suffered crushing losses from the Egyptian advance. The Egyptians had crossed the canal with massive numbers of *Sagger* anti-tank guided missiles (ATGMs) and rocket propelled grenades (RPGs) to compensate for the lack of armor in the first wave of the advancing troops. Even they were surprised at how well this tactic worked, destroying two-thirds of an Israeli armored division in less than 1 day.[56] Dayan subsequently gave the order that the southern front commander was to stop wasting his forces trying to save the soldiers in the besieged forts. Rather, the strength of the Army had to be preserved, and the men from the forts had to make their way east as best they could.

As expected, the Egyptians used their older *Tupolev* Tu-16 (*Badger*) bombers to fire AS-5 *Kelt* missiles at targets deep in Sinai on the first day of the conflict. One *Kelt* was also fired at Tel Aviv on the same day, but was shot down by an Israeli fighter aircraft.[57] The Egyptians apparently wanted to underscore their ability to strike Israeli homeland targets with long-range weapons, and only a limited use of *Kelts* was necessary for that purpose. These large and slow missiles were originally designed as anti-shipping weapons and carried a massive 2,205 lbs. conventional warhead. They have a range of around 120 miles. The Tu-16 sorties with *Kelts* resulted in two direct hits on Sinai radar sites, totally obliterating them. The strikes may have been a signal to the Israelis of the Egyptian ability to launch attacks into the Israeli hinterland, although they would not have been much more than terror weapons if they remained armed with conventional warheads.[58] If these attacks were meant to convey an implicit threat of possible chemical or biological attacks at a later time, they presented something of a mixed message since 20 other Egyptian *Kelts* were shot down by Israeli aircraft or anti-aircraft fire before they were able to strike targets in Sinai.[59] The earlier criticisms of the *Kelt* put forward by General Bagdadi were borne out by these incidents. The decision to fire a *Kelt* missile at Tel Aviv seems quite provocative, and was not repeated during the remaining weeks of the war.

Key Israeli military leaders were surprised, if not

shocked, at the Egyptian accomplishment of crossing the Suez Canal and the equally unexpected fighting spirit of both the Egyptian and Syrian forces. Nevertheless, many of these same leaders still harbored strong doubts about sustained Arab military effectiveness and expected that Israel would quickly gain the initiative and recover from the situation quickly after Israeli reserves had been mobilized.[60] Defense Minister Dayan was something of an exception to this mindset and was more willing to question the conventional wisdom. In an October 7 afternoon meeting with Prime Minister Golda Meir and other Israeli leaders, he raised the possibility of withdrawing to a new defensive line deeper in the Sinai.[61] He opposed an offensive on the southern front and stated that he did not believe that the Egyptians could be pushed back over the Canal even when fresh Israeli forces from the reserves were placed into the fight. He also stated that, "This is a war for the Third Temple [the survival of Israel], not for Sinai."[62] Dayan concluded that all previous Israeli strategic assessments were based on the experience of the June 1967 War and were without much value in planning future courses of action.[63] According to Dayan's interpretation, Israel was fighting for its survival and was losing. Others at the meeting, including IDF Chief of Staff Lieutenant General David Elazar, felt Dayan was overly pessimistic. Elazar later stated that Israeli Prime Minister Golda Meir privately told him that if Dayan was correct, her "entire world" would collapse and that she would have no reason to go on living.[64]

Most of the senior military officials that met with Prime Minister Meir on October 7 favored a military offensive in Sinai as soon as they were able to mobilize adequate numbers of reservists. Israel needed a quick decisive victory rather than a long war of attrition that could bankrupt the state. At this point, few of the Israeli leaders, except for Dayan, believed that the Egyptian army had changed so fundamentally that it might be able to outfight the IDF after Israeli reserve units entered the battle. At that point, Israel would have sufficient numbers of troops to exploit its perceived advantages in tactics, training, and fighting spirit. In public, Defense Minister Dayan put aside his doubts and supported this viewpoint when he told Israeli newspaper editors, "The Egyptians have embarked on a very big adventure they haven't thought through. After tomorrow afternoon [when the reserves enter the battle], I wouldn't want to be in their shoes."[65] Dayan hoped such a statement would enhance military and public morale and also wanted to project enough optimism to discourage other Arab states from entering the war. On the southern front, plans were made for a counterattack using newly mobilized forces to begin on the morning of October 8. This was to be arguably the worst day in Israeli military history and one that may have pushed the Israelis to consider previously unthinkable options.

As the October 8 counteroffensive unfolded, Israeli tank units attacked into the fortified strong points of the Egyptian bridgehead with insufficient levels of artillery support, air cover, and infantry support. To make matters worse, the tanks often surged ahead of their limited infantry support, much of which was in older and slower World War II type half tracks. These forces were consequently highly vulnerable to Egyptian soldiers using ATGMs and RPGs.[66] Unlike the 1967 war, Egyptian infantrymen stood their ground and inflicted massive losses on the Israeli units, forcing them to retreat in disarray. Major General Shmuel Gonen, the

Commander of the Sinai Front at the outset of the war, commented, "It's not the Egyptian Army of 1967."[67]

At the time, the Israeli counterattack of October 8 was ferociously criticized by Major General Ariel Sharon, a former leader of Israel's Southern Command recalled to active service. Sharon believed that the Israeli forces were too small and poorly deployed to push back the Egyptians. He also saw no need for an immediate counterattack stating that "The Egyptians aren't headed for Tel Aviv . . . They can't afford to go beyond their missile cover."[68] Conversely, General Gonen did not seem to grasp the strategic situation and was widely viewed to be out of his depth in the ongoing struggle.[69] Eventually, the Israeli leadership recalled former Chief of Staff Lieutenant General Chaim Bar-Lev (then serving as Commerce and Industry Minister) to active duty and placed him in command of the Sinai front over both Gonen and Sharon.

Events on the northern front were better for the Israelis on October 8 but far from under control. The war had begun on October 6 when Syrian commandos inserted by helicopter seized key terrain on Mount Hermon capturing a key Israeli intelligence outpost with all of its sensitive surveillance equipment. The Israeli 7th Armored and 188th (Barak) Brigades were almost totally destroyed by advancing Syrian troops. By October 7, Major General Dan Laner of Northern Command stated "The War in the south[ern Golan] is over, and we have lost."[70] This statement was an exaggeration, but it reflected the extremely tough fighting at the time. On October 8, Israel's elite Golani Brigade also failed to retake Mount Hermon as planned. This failure occurred partially because Syrians had better night fighting equipment.[71] The Syrian advance was finally halted on October 9 due to the

Israel decision to give that front priority for reenforcements and especially air support missions.

The massive Israeli defeat in Sinai on October 8, along with the continuing problem of the Golan Heights, led to what one source referred to as a "seemingly inexorable spiral to defeat."[72] In the first 48 hours of war, 400 Israeli tanks were lost in Sinai alone, a staggering and jolting setback.[73] In the first 2 days of the war, Israel lost a number of combat aircraft, with little to show for these losses.[74] While Egypt, at this time, was not moving its forces beyond the protected beachhead, serious questions of the Israeli military's long-term viability remained. Defense Minister Dayan was particularly concerned that Israeli combat resources, (particularly tanks and aircraft) were being destroyed so quickly that Israel could not continue to maintain the tempo of the conventional war. Moreover, the deployment of large numbers of reserves over an extended period of time removed productive individuals from the Israeli civilian work force while dramatically escalating government spending. Since almost all qualified Israeli men were then required to serve in the reserves until age 55, a full mobilization, such as occurred in 1973, could eventually cause the economy to collapse. For the Israelis, a prolonged battlefield stalemate was simply another path to military defeat.

Israel's October 8 defeat in Sinai raised the possibility that additional Israeli problems might result from a regional perception that Israel was losing the war. The Israelis had to worry that additional Arab countries would enter the war if the fighting continued for a significant period of time. Two countries of particular concern in this regard were Iraq and Jordan.[75] Jordan, which had been badly mauled in the 1967 war, was not interested in doing anything that could invite an Israeli conventional strike against Jordanian targets. King Hussein was, however, under intense pressure to either join the battle or open his territory to Palestinian guerrillas so that they could strike into Israel.[76] Syria made a formal request for Iraqi military assistance on October 10. Baghdad agreed to this request on the same day. Also on October 10, the Jordanian leadership called up their reserve forces in anticipation of playing a military role in the conflict. While Jordan did not open a third front against Israel, it was eventually to send the 40th Armored Brigade to the Golan Heights, where it fought with distinction. The 40th did, however, have considerable difficulty coordinating with the Iraqi forces that were also involved in the fighting thereby harming the effectiveness of both forces.[77]

The Challenge to Israeli Decision-Makers and the Potential for Israel to Escalate the War.

Defense Minister Dayan held a special role in Israeli society on the eve of the attack. As IDF Chief of Staff in the 1956 War and defense minister in the 1967 War, he was associated with some of Israel's greatest victories. Dayan became Chief of Staff in December 1953 at age 38. He remained in that post until 1958 when he retired from the army and entered politics. He then became defense minister in May 1967 on the eve of the Six Day War and remained in that position until 1974. As defense minister in the 1967 War, Dayan ordered the seizure of the Golan Heights on his own authority (without consulting the prime minister) in the final days of the fighting. In 1973, most Israelis and many foreign experts considered Dayan to be a tough, aggressive, military genius tempered by decades of experience commanding a victorious army. No one's military judgment was more respected.

In contrast to Defense Minister Dayan, Prime Minister Meir had no military background and no knowledge of military subjects. In a conversation with her military aide, Major General Yisrael Lior, she confessed that she had no idea what a division was.[78] She was therefore extremely reluctant to overrule the judgment of military professionals. Prime Minster Meir, then 75 years old, was described as surviving on coffee and cigarettes throughout the 3 weeks of crisis and war.[79] At this stage in her life, she also had a number of health problems. Consequently, she was required to make a number of deeply challenging decisions under extremely difficult circumstances, while receiving highly negative interpretations of major events from her defense minister. While both Meir and Dayan were shocked at the newfound military effectiveness of the Arab armies, Dayan was the most shocked. Moreover, his vast military experience suggested that his pessimism was solidly based on an in-depth understanding of the military dangers Israel faced from neighboring Arab states. Prime Minister Meir had no reason to doubt Dayan on his military judgments or to assume that his pessimism was based on anything other than a superior and more insightful understanding of the military situation than she could ever possess.

In April 1976, *Time* magazine ran a special report entitled, "How Israel Got the Bomb."[80] This was the first major published source indicating that Israel considered using nuclear weapons in the 1973 War. A number of non-Israeli secondary sources have confirmed the *Time* magazine story, often adding additional details from undivulged sources. Journalist Howard Blum gives a particularly detailed discussion in his 2004 book, *The Eve of Destruction*. An earlier study by investigative journalist Seymour Hersh also provides an in-depth discussion of these events. Israeli scholar Avner Cohen has probably gone further than any other Israeli intellectual in publishing on these nuclear issues, but unlike the Americans, Cohen had to be wary of censorship laws that could become a problem for him as an Israeli citizen.[81] Another extremely valuable study by an Israeli scholar with access to the top Israeli decision-makers at the time has been provided by Shlomo Aronson.[82] All of these sources agree that Israel had nuclear weapons and was considering using them. If so, the weapons would not have been previously subjected to nuclear testing, but the Israelis may have had enough confidence in the design of their weapons to have confidence in them regardless of this situation.

Dayan appeared to be overwhelmed by the events of October 8 and especially the massive Israeli defeat in Sinai. According to Major General (later president of Israel) Chaim Herzog, Dayan went to Prime Minister Meir on October 9 and told her, "Golda, I was wrong in everything. We are headed toward a catastrophe. We shall have to withdraw on the Golan Heights to the edge of the escarpment overlooking the valley and in the south in Sinai to the passes and hold out to the last bullet."[83] The *Time* story indicated that on October 9, Dayan stated, "This is the end of the Third Temple" by which he meant the state of Israel. According to this version of events, Dayan suggested the use of nuclear weapons might be required to save the country. By this time, the Israeli leaders had become aware that Egypt was conducting a limited war and that Egyptian troops were then entrenching themselves in their positions near the Canal. For the time being, that problem did not seem to require a nuclear response. Syria was another matter. If the Syrians were able to crush Israeli defense on the Golan Heights, it appeared possible that they could strike into Israel itself. While post-war analysis was to indicate this was unlikely, no one was prepared to appear to appear too smug about the potential of Arab armies on October 9. Too many illusions had been shattered, and too many people had been wrong. Some sources also suggest that Israel was prepared to use nuclear weapons against the Defense Ministries of Cairo and Damascus.[84]

Attempting to reconstruct Israeli decision-making in the first week of the war, Israeli scholar Shlomo Aronson asserted, "Yigal Allon, Dayan's old enemy, whispered loudly behind the scenes (including in a conversation with me in 1976) that the minister of defense had indeed intended *to use nuclear weapons against the Syrians* [his italics]."[85] Aronson asked Dayan about this statement, and the former defense minister absolutely denied it.[86] Dayan's friend and former protégé, Shimon Peres, further dismissed the allegation as meaningless by noting that early in the war a variety of options were discussed and that discussion does not equate to a decision to move on a particular option. Aronson concludes that a great deal of spite went into the Allon allegations, and that he was attempting to besmirch the reputation of his political (and personal?) enemy.[87] Yet, no one involved in this discussion seems to be denying that Israel has nuclear weapons in 1973, and there was at least a discussion of what to do with them.

In a closed 1996 lecture that was disclosed by the Israeli press a year later, nuclear physicist and former Israeli Science Minister Yuval Ne'eman stated the Israeli leadership ordered the deployment of its *Jericho I* surface-to-surface missiles without camouflage as a signal to the Arab militaries that Israel would use these systems under certain circumstances.[88] He stated that no indication was given as to whether the warheads for these missiles were equipped with conventional or special types of munitions, and the Soviets and Arabs were left "guessing what type of Israeli warheads" had been emplaced on them. It did not, however, make much sense for Israel to have an expensive 500 kilometer range delivery system, such as the *Jericho I*, used to deliver a small quantity of high explosive ordinance. For the weapon to have any strategic value, it had to carry some sort of WMD. This payload would most likely be a nuclear weapon, although a biological weapon is a militarily reasonable alternative if the Israelis had chosen to develop and weaponize biological weapons.[89] The deployment of the missiles may have may have been a signal to the Arab countries, the Soviet Union, and perhaps the United States that there were limits beyond which Israel was not prepared to be pushed before these strategic systems came into play. The Israeli leadership probably assumed that U.S. and Soviet intelligence agencies would detect the new deployments, and that the Soviets would communicate this information to Cairo and Damascus. There are also reports that Dayan asked Secretary of State Kissinger to pass a nuclear warning to Damascus.[90]

Israeli historian Martin van Creveld has also speculated that the Israeli deployment might have been "a veiled Israeli hint" to the Syrians that certain conventional actions on their part could produce a nuclear response.[91] He further suggested that the Syrians may have had some military opportunity to push forward into Israeli territory or use their heliborne forces to seize the Israeli bridges over the Jordan River and thereby achieve some potential to both block Israeli resupply and to open the door to their own forces for movement deep into Israeli territory. Van

Creveld implies that the Syrians were deterred from these courses of action. He then quotes Syrian Defense Minister Mustafa Tlas who claims that a deliberate decision was made not to press into Israeli territory, but the time to discuss the reasons for this decision had not yet arrived.[92]

Another Israeli scholar, Yair Evron, stated that during this time frame, Israel did not prepare its nuclear weapons for use despite numerous reports to the contrary. According to what Evron refers to as "reliable accounts," Dayan's interest in potential nuclear weapons use was stringently opposed by two of Meir's closest advisors, Minister without Portfolio Yisrael Galili and Deputy Prime Minister Yigal Allon, a former IDF Chief of Staff.[93] Additionally, Prime Minister Meir increasingly relied on the then-Chief of Staff, Lieutenant General Elazar, rather than Dayan for military advice. Although Elazar had made numerous pre-war mistakes that were now becoming apparent, he remained calm in the midst of the crisis. Dayan, by contrast, admitted that he was gripped by an anxiety that he had never previously experienced.[94] He, nevertheless, continued to show reasonable strategic judgment in public throughout the war, although it was often mixed with unrelenting pessimism that critics say verged on defeatism. Dayan responded to these critics by noting that some of his suggestions to the Chief of Staff were drastic, but they were also firmly based in reality. Dayan and his supporters later claimed that he was a "constructive pessimist."[95] Dayan further implied that Prime Minister Meir was more distressed than he was at times because she sought poisons to be able to commit suicide in the event that Israel was overrun.[96] Dayan attempted to contrast that approach to his own behavior, stating "Everybody praised during the war the strong spirit of the Prime Minister, the late Mrs. Meir. And I think very much of her. I admire her. Still you know that she thought about committing suicide, Mrs. Meir. It never occurred to me. Absolutely never occurred to me. My spirit was never broken."[97] Dayan's criticism is somewhat unfair since Meir would only consider such an option if Israel was defeated and occupied, and Dayan's comments had been a major factor convincing her that this destruction was possible.

Investigative journalist Seymour Hersh maintains that Israeli officials told the U.S. leadership on October 8 that their country was prepared to use nuclear weapons against the Arab states unless the United States initiated a massive resupply effort for conventional weapons and munitions.[98] Such a "nuclear blackmail" approach seems somewhat farfetched, despite the desperate circumstances, because it was almost certainly unnecessary. While the United States did engage in such an effort (Operation NICKEL GRASS) which started in the second week of the war, it is by no means clear that this operation was a response to Israeli threats to use nuclear weapons against the Arab states. An equally likely motivation was the initiation of a Soviet weapons and supply airlift to Egypt and Syria on October 9. Moscow's airlift was limited during the first week of the fighting but expanded dramatically on October 12. Many of these flights seem to have been directed toward Syria to replace its massive equipment losses in the first 5 days of the war.[99] The United States would have had great difficulty ignoring the dramatic increase in Soviet involvement in supplying the Arab states with massive wartime assistance, and a strong U.S. countermove was virtually certain within the logic of Cold War competition.

The nuclear blackmail scenario may also assume

that the United States was prepared to place serious limitations on the resupply effort without an Israeli threat of nuclear weapons use against the Arabs. This belief seems to be based on the slow and limited level of the U.S. resupply effort at the beginning of the war, as well as exaggerated accounts of conflicts between U.S. cabinet officials over the level of aid that would be appropriate.[100] It is true that at the outset of the fighting, the effort to resupply Israel began with only a small semi-covert effort in which Israeli El Al passenger aircraft and charter aircraft carried weapons and equipment to Israel, but these actions were initially viewed as adequate because of the stereotype of massive Israeli military superiority. As a hedge, to ensure that Israel's powerful air force was bolstered with ongoing support, U.S. Navy *Skyhawk* aircraft were also flown to Israel from Europe, transferred to Israel, and then repainted with the Star of David for immediate use in combat.[101] Until at least October 8, most U.S. leaders, including President Richard Nixon, expected a lightning Israeli victory as occurred in 1967. Such an outcome was also predicted by the U.S. intelligence community and by the Israelis.[102] Immediately, after the war broke out, Israeli Deputy Ambassador to the United States Mordechai Shalev told Secretary of State Kissinger, "You shouldn't worry."[103] Only after the full extent of Israeli military setbacks became known was the U.S. Air Force ordered to undertake what became one of the most massive resupply efforts in aviation history. Operation NICKEL GRASS began on October 13, with the first of many U.S. Military Airlift Command (MAC) aircraft arriving in Israel on October 14.

Additional Military Developments Influencing Israeli Decision-Making and Israeli Conventional Strikes Against Syrian Strategic Targets.

While Israeli nuclear weapons were not used against Syrian targets, the Israelis did dramatically escalate the conflict beyond the battlefield in response to Syrian use of Free- Rocket-Over-Ground-7 (FROG-7) rockets early in the war. On October 7 and October 8, the Syrians fired FROG-7 rockets which struck throughout central Galilee in an attack directed at Israeli air bases in the area including Ramat David.[104] As implied by the name, FROG-7s have no internal guidance system and are targeted through the use of END TRAY meteorological radar. Pinpoint accuracy was never a priority for Soviet designers since the FROG-7 (like the *Scud*) was to be used as a nuclear or chemical delivery system for the Soviet Army against area targets. Problems with accuracy, endemic to this weapons system, were dramatically reflected in Syria's 1973 attacks which were the first time that FROG-7s were used in combat. Syrian rocket crews might also have been inadequately trained and could not overcome the rockets' inherent problems with accuracy. Consequently, the FROG-7 attacks failed to hit any of the Israeli air bases and instead appeared to be indiscriminate terror attacks against civilian settlements in northern Israel. While Israeli military intelligence understood the situation, the Israeli military moved quickly to exploit the Syrian ineptness with rockets.

An October 9 Israeli decision to bomb important targets throughout Syria was publicly presented as a response to the rocket attacks in Galilee. This claim helped to justify the decision to escalate the fighting to civilian targets and thereby minimized the possibility for diplomatic friction with the United States.[105] This strategic bombing campaign was directed against Syrian infrastructure targets including port facilities, storage depots, storage facilities for fuel, power plants, and industrial facilities. On October 10, the IAF bombed the Syrian Defense Ministry in Damascus, although only limited damage was inflicted on the defense facilities.[106] The Soviet Cultural Center in Damascus had previously been bombed on October 9, and some Soviet diplomats were killed. Later, on October 12, the *Ilya Mechnikov*, a Soviet merchant ship, was sunk by Israeli aircraft attacking the Syrian port of Tartus.[107] The Israelis apologized for the deaths of Soviet citizens, but Soviet UN Ambassador Malik only used the apology as an occasion for crude and highly theatrical diplomatic abuse. Soviet Foreign Minister Andrei Gromyko considered Malik's behavior excessive, referring to the episode as "a cheap show," and instructing him to show more restraint.[108]

One explanation for the Israeli bombing campaign that was popular with both Western and Arab analysts involves a belief that the Israelis had become emotional and frustrated over the early Arab victories. According to Egyptian author Mohamed Heikal, "Frustrated by the unexpected resistance of the Syrian troops . . . and by the scale of his own losses the enemy turned the night of October 9 into an inferno while the air force kept up continuous strikes."[109] Also according to Heikal, "Damascus and Homs were heavily bombed in an attempt to weaken civilian morale and provoke a collapse of the regime." Yet the simplistic explanations for the attacks offered by Heikal may have masked more sophisticated reasons for the attacks.

Beyond signaling or even revenge, there may have been important operational reasons for the Israeli bombing campaign against Syria. These operational concerns centered on Israeli plans to defeat Syria quickly so that they could then turn all of their attention to the Egyptian front. Air strikes throughout the country could play an important role in meeting this objective by disrupting Syrian logistics and weakening morale at the front. The Israelis might have also hoped to pressure the Syrians into redeploying some of their air defense assets to protect infrastructure targets, thus diverting or even drawing them away from the front lines. By October 11, the IAF had already destroyed enough SAMs on the northern front to seriously weaken the integrated nature of the air defense system.[110] This change allowed the IAF to increase the intensity of close air support to ground units, dramatically enhancing their ability to achieve battlefield success.

Egypt, for its part, was unwilling to escalate the war for the sake of its northern ally. Sadat rejected a Syrian request that the Egyptian Air Force undertake deep raids against Israel in response to the attacks, according to senior Soviet diplomat Victor Israelyan.[111] Cairo was already taking a number of serious risks and was unwilling to let the war escalate to a point where Egypt's strategic goals were threatened. If one of the Israeli objectives for the bombing raids was to drive a wedge between Cairo and Damascus, they were clearly successful.

The Turning of the Tide and the Potential for Arab and Soviet Escalation of the War.

If the Israelis had seriously considered the use of nuclear weapons after their October 8 defeat, the pressure for them to consider such options slowly receded after that date. The October 8 Israeli defeat in Sinai was followed by a series of successful Israeli operations on the northern front in which Syrian troops were pushed off the Golan Heights. The fighting on the Golan Heights gave the Israelis the opportunity to excel at the type of warfare that they had consistently stressed in their training and doctrine, tank-to-tank combat. The loss of increasing numbers of Syrian anti-aircraft weapons supporting the ground troops eased the task of IAF close air support, while Jordanian and Iraqi forces that had rushed to fill the gap had trouble not only communicating with each other but also had serious difficulties coordinating their military operations with the Syrians. At least six Iraqi jet aircraft sent to support the Arab effort were shot down by Syrian missile crews who identified the aircraft as hostile.[112] On October 11, Prime Minister Meir agreed to a limited offensive to place the Israeli army close enough to Damascus to threaten the capital with long-range artillery. By October 13, the Israelis had achieved this objective, and artillery fire from 175 mm guns was directed on Damascus Airport, just southeast of the city. Intermittent artillery fire against the airport continued for the next 10 days.[113]

A reversal of Israeli fortunes in the Sinai occurred shortly after their victories on the northern front. On October 14, Egyptian forces launched a major offensive in the Sinai in an effort to reach the strategic Mitla and Giddi passes about 30 miles east of the Suez Canal. The chief motivations for this action were based on political imperatives rather than military strategy. The Syrian military had suffered such serious military setbacks that Damascus was in danger of being knocked out of the war. The offensive push on the southern front was designed in part to relieve this pressure on Syria and to indicate an Egyptian response to the bombing of Damascus and other urban centers. President Sadat pushed forward with this operation overruling the objections of Lieutenant General Shazly, who strongly maintained the Egyptian troops should not move forward and out of the protection of the air defense umbrella over the Canal.[114] Serious military planning and training for this type of operation had not occurred, unlike the meticulous planning and preparation that characterized the effort to cross the Suez Canal.[115] The Egyptians were correspondingly setting themselves up for disaster.

The Egyptian troops advanced under the protection of some mobile air defense systems but left the protection of their carefully constructed network of Canal air defenses, which integrated a variety of systems with different ranges and capabilities. Air defense weapons are optimal only if employed in an interlocking network where the ranges and capabilities of different systems are used to support each other and deny hostile aircraft any undefended airspace. By leaving the Canal defenses, the Egyptians exposed themselves to the full fury of the IAF which quickly exploited gaps in the protection for advancing troops. While the advancing Egyptian troops had air defense assets, they no longer had an air defense system. This deficiency helped lead to a serious Egyptian defeat. In the resulting battle, the Egyptians lost almost 250 tanks and 200 other armored vehicles, while the Israelis lost no more than 40 tanks.[116] The Israeli victory led to an immediate improvement in IDF morale. On the evening of October 14, Lieutenant General Bar-Lev told the Prime Minister Meir, "We are back to being ourselves, and they are back to being themselves."[117]

The war continued to go badly for Egypt when Israeli military units crossed the Canal into Egypt in force on late October 15 and 16. Forces under the command of Ariel Sharon led the crossing and then were dispatched to raid Egyptian positions on the west bank of the Canal. The Israeli forces attacked a number of targets including garrison forces, fuel depots, some tanks, and most importantly a number of SAM and AAA sites. The destruction of these sites opened up a 15-mile gap in Egypt's air space. The Israeli position on the Canal's west bank was correspondingly defended with air strikes which could now be freely employed to support the ground troops. As additional Israeli ground units crossed to the west bank, the IDF was able to expand the Israeli enclave using the integrated armor and air tactics at which they excelled.

Egyptian Attempts to Contain Israeli Military Gains and the 1973 *Scud* Launches.

The Egyptians had enjoyed a level of military success that not even they had expected until October 14 when their situation turned around dramatically. The Syrians, as noted, began experiencing serious military setbacks much earlier in the conflict. Prior to October 14, the dominant escalatory danger was that Israel would become increasingly desperate and consider the use of nonconventional weapons to reverse a tide of Arab military success. After October 14, the Israelis fought a number of extremely difficult battles, but they also made steady progress toward military victory. The threat of Israeli escalation to unconventional weapons correspondingly receded, while the danger of Arab escalation, to the extent this was possible, became more serious. In particular, Cairo increasingly sought to find ways to end the war on terms that allowed it to preserve some of the Egyptian gains made early in the war, while preventing outright defeat. The Egyptian leadership wanted a cease-fire and was prepared to issue tough threats to help achieve this objective.

Sadat gave his most important speech of the war on October 16 after the Egyptian defeat 2 days earlier. By this time, Israeli troops had already crossed the Suez Canal, but the Egyptian president may not have realized that this development had taken place. He did know, however, that Egypt was facing an extremely serious military challenge. In this speech to the Egyptian parliament, Sadat quickly asserted that he would agree to a cease-fire with Israel if it returned to the June 1967 borders. He further stated that Egypt would then be willing to attend a peace conference with Israel, and that it would encourage the other Arab states to do so as well. While such terms were hardly generous in light of the military situation at the time, Sadat was clearly signaling an interest in permanent peace which different audiences responded to in different ways. This speech unsurprisingly did nothing to reassure the Israelis during the ongoing fighting, but it deeply upset Egypt's Syrian allies, who felt that Sadat was making important decisions without consulting them.[118] More ominously, Sadat also issued a warning in the October 16 speech when he claimed Egyptian missiles were "on their pads, ready with one signal to be fired into the depth of Israel."[119] Sadat stated that Egypt's German-designed *al-Kahir* missiles would be used against Israel if the Israelis chose to bomb Egyptian cities, as had occurred with the bombing of Damascus. Since the *al-Kahir* was not a viable weapon, it was widely assumed that he was hinting at potential *Scud* use. Sadat would eventually make good on this threat to fire *Scuds* at the Israelis, although they would have only conventional warheads.

Egypt was the first country to use *Scuds* in combat when it did so in the 1973 War. At least two and possibly three *Scuds* were fired on October 22 with the military objective of destroying one of the Israeli pontoon bridges on the Suez Canal. These bridges were extremely important military targets, but a *Scud* missile with a conventional warhead is a particularly poor choice of a weapons system for a pinpoint target. The *Scud's* primitive inertial guidance system was completely inadequate for such a strike, and the Israeli bridge was not hit. The missiles, nevertheless, appear to have been launched with a larger agenda than seeking to destroy the bridge. Sadat stated, "I wanted Israel to learn that such a weapon was indeed in our hands, and that we could use it at a later stage of the war; even though Israel had in fact realized from the moment the war broke out that we meant and did what we said."[120]

The *Scud* missiles themselves were of almost no strategic value without a more threatening warhead. In this regard, a number of important sources have suggested that Sadat now felt it was important to have more options than simply a vague and untested biological weapons capability. The same *Time* special report that broke the story of Israeli nuclear options in the war also stated, "What is certain is that on October 13, the Russians dispatched nuclear warheads from Nikolaev—the naval base at Odessa—to Alexandria, to be fitted on Russian *Scud* missiles already based in Egypt."[121] While this report seems to be mistaken about the ownership of the *Scud* missiles, it is, of course, important. Some other interesting developments seem to relate to this incident. William B. Quandt, a staff member on the National Security Council during the war, has commented on what appears to be a second such incident when he notes that a suspicious Soviet freighter passed through the Bosporus on October 22 and arrived in Alexandria on October 25.[122] According to press conferences by both Secretary of State Kissinger and Secretary of Defense James R. Schlesinger, this ship was believed to be carrying nuclear weapons.[123] Quandt suggests that the Soviets were, by this time, deeply committed to supporting their allies by strengthening the shaky cease-fire and also raised the alert status of some of their airborne forces. He goes on to comment, "Whether [the activities of the freighter were] intended as some form of signal to the United States that the Russians would not allow the Egyptians to be defeated by the Israelis is unknown. . . . [P]erhaps the Russians, expecting the Americans to monitor the ship's nuclear cargo, were engaging in psychological warfare aimed at convincing Washington of the need for an early end to the fighting."[124]

In an article dated November 5, 1973, but probably released before that time, the respected aerospace magazine, *Aviation Week and Space Technology*, carried the remarkable statement that, "Two brigades of Soviet Scud surface-to-surface missiles, each equipped with a nuclear warhead, are now in a position east of Cairo, poised to strike Tel Aviv and other Israeli population centers."[125] Also according to *Aviation Week*, "The Russian's made little if any effort to hide from the U.S. the fact that nuclear warheads had been shipped to the Middle East." The warheads were reported to be uncamouflaged and resting near the Egyptian *Scud* launchers. If true, the Soviets would probably only consider allowing these warheads to be used in response to an Israeli nuclear attack. Yet the Israelis had already essentially achieved their key military objectives by the time the suspected warheads became an issue. This situation suggests that Quandt's psychological warfare explanation is especially plausible.

As their military situation continued to deteriorate,

the Egyptians became more deeply concerned about implementing a lasting cease-fire. On October 23, a UN sponsored cease-fire went into effect, but it rapidly broke down, with Israel benefiting from the collapse. Israeli Major General Avraham Adan's division fought its way south against tough opposition and reached Suez City on the evening of October 24. The city was captured after heavy fighting. As a result of this action, the Israelis had almost completely encircled Egypt's 40,000-man Third Army which was then in Sinai. Israel was then in the position where it could prevent the resupply of the Third Army and thereby force its commander to surrender unless he could break the encirclement. The Egyptian Second Army may also have been placed in danger of encirclement if the war had continued. Under these dire conditions for Egypt, the cease-fire was reestablished on October 25 after intense diplomatic activity by U.S. Secretary of State Kissinger. The war was finally over, and Kissinger did everything he could do to stabilize the cease-fire and thereby avoid humiliating Egypt by forcing a surrender of massive numbers of troops in the field. Egypt had only avoided catastrophic defeat through the diplomatic intervention of the United States. The U.S. leadership had taken a chance on Sadat who had been using secret channels to indicate his willingness to break with the Soviet Union and seek a new and closer relationship with the United States that would help to facilitate a lasting Arab-Israeli peace.

The 1973 War is a particularly interesting and informative case study because it involved one combatant (Israel) and an opposing (Arab) coalition, which, while not equal in strength, were at least less unequal than in the case study that follows on the 1991 Gulf War (between the U.S.-led coalition and Iraq). The lessons of escalation in the 1973 case study are discussed in some depth in the conclusion of this work, but the 19-day war clearly underscores how rapidly evolving conventional combat can lead to escalating and receding dangers of escalation to WMD use as well as the strong dangers of escalation brought on by misperceptions of the adversary's intentions.

THE 1991 IRAQ WAR

Deterrence and Escalation Control.

The 1991 Gulf War is an a second example of a conflict involving the danger of uncontrolled escalation and the use of intrawar deterrence. It was, however, a very different type of war, and Iraq maintained a vastly dissimilar type of political and military system from the Israeli government that was forced to make critical decisions on escalation in 1973. The two wars were also completely different types of confrontations. In particular, the 1973 war involved efforts by Egyptian and Syrian forces to make territorial gains and thereby achieve a limited victory, while simultaneously avoiding pushing Israel to use of its suspected nuclear weapons. Conversely, in the build-up to the 1991 conflict, Saddam Hussein had already seized the territory that he was interested in retaining. His goal was to maintain control of Kuwait and, failing that, to at all costs remain in power in Iraq. Saddam Hussein thus sought to retain Kuwait, while deterring an attack by the U.S.-led coalition if at all possible. If an attack occurred despite his best efforts to prevent it, his core goal then became regime survival which was key and non-negotiable.

To understand Iraqi actions during the 1991 Gulf War, it is necessary to consider the nature of the Saddam Hussein regime. A number of observers have noted that it is easier to seize power in Iraq than to retain it. Another complication is that coups in Iraq have seldom involved those individuals who lose power quietly going into exile. Rather, past Iraqi coups were often bloody, and leaders of a deposed regime are not allowed to live to fight another day if this can be prevented. Saddam eventually took power after a long series of Iraqi coups beginning with the overthrow of the monarchy in 1958. Unlike his predecessors, Saddam was able to establish a stable authoritarian regime by creating a regime so regimented and brutal that other Iraqi leaders did not, under normal circumstances, have the political space to organize a successful effort to overthrow the dictator. The skill he displayed with more proficiency than any other Iraqi leader was the ability to stay in power. Saddam remained the leader of what often appeared to be a coup-proof regime until he was deposed by the United States in 2003.

Saddam's regime did not maintain any constraints on the dictator's power, and his disfavor could result from some innocuous misstep on the part of his subordinates or their failure to perform to his expectations. Saddam generally frightened everyone around him, and to irritate him in any way was to court disgrace, imprisonment, torture, and execution. As with virtually all authoritarian leaders, he valued loyalty above any other virtue. Speaking unpleasant truth to power was seldom acceptable in Iraq because it could be interpreted as questioning the dictator's authority by pointing out flawed decision-making and thereby displaying the limits of one's loyalty. Instead, Saddam's subordinates often watched for indications of their leader's opinions on a subject and then fell all over themselves to agree with him. One Iraqi general characterized this approach as consistent with long-standing Iraqi views on acceptable social behavior. According to General Raad Hamdani, who commanded Iraq's Second Republican Guard Corps in early 2003, "The truth is, in our culture, frankness is disrespect."[126] No one living under the Iraqi regime wanted to show any disrespect to Saddam.

Yet, there was some also some danger that Iraqi leaders who failed to warn Saddam of potential problems could find themselves accused of duplicity and incompetence once the dictator became cognizant of any problems that had previously been minimized. During the Iran-Iraq war, military intelligence officials were in special danger of his attributing military setbacks to a lack of "correct information."[127] Thus, Saddam's subordinates often had to walk a tightrope between some minimal actions to prepare the dictator for bad news and an effort not to offend or upset him with unpleasant realities that had resulted from his earlier political and military decisions. A further complication was that Saddam was not always consistent in the ways he dealt with subordinates. According to General Handani, "Sometimes you would feel so close to him that you could spill your heart to him, and other times you felt like you were in a cage with a hungry lion."[128] During the 1991 war, some officials tried to sidestep the dangers of too much honesty by noting reports by the international media that pointed out Iraqi setbacks and shortcomings. Saddam tended to dismiss such media reporting as "lies" and "psychological warfare." In those rare cases where a subordinate was able to bring himself to carefully and respectfully disagree with Saddam's judgment, the subordinate was seldom able to change the dictator's mind and would never be foolish enough to pursue the topic once he had dismissed it.

Saddam Hussein's judgment and thinking may also have been shaped by a belief in his own genius.

The ability to rise from a deeply impoverished childhood to the undisputed leadership of Iraq is a stunning accomplishment, and this kind of success can bolster fantasies about one's own infallibility. Additionally, Iraq, under Saddam's dictatorship had one of the most elaborate cults of personality in the world under which he and almost everyone else in the country was bombarded with continuing propaganda about such topics as the dictator's brilliant mind and impeccable judgment. Saddam's increasing recklessness, which he may have seen as boldness, probably grew out of excessive self-confidence in his ability to deal with any situation that his country faced. The Iraqi dictator further considered himself to be a master of manipulation and deception and sometimes had difficulty accepting fairly frank statements by foreign and particularly hostile states at face value. Straightforward messages were most often scrutinized for hidden meanings or attempts at misdirection were always to be considered.

Despite all of these problems, it might also be noted that Saddam did not lose control of his emotions and become irrational.[129] While he was an exceptionally violent man, his casual use of violence was always part of a larger set of concerns. Saddam became angry on a number of occasions, but he did not kill people out of passion (as his eldest son, Uday, did, much to his father's contempt). Rather, Saddam controlled his passions and used cold and determined violence, torture, and execution to ensure that his goals for Iraq were properly served and that he remained Iraq's unchallenged and unchallengeable leader. According to one Iraqi scholar, "The pent-up violence in [Saddam's] personal makeup was always controlled and directed by a political sense of judgment . . . Saddam [Hussein], unlike Idi Amin or Papa Doc, is marked by this calculated, disciplined, and above all effortless resort to violence genuinely conceived to be in the service of more exalted aims."[130] Thus, cruelty and murder were used to support the regime's longevity and to ensure its goals were met without challenge or question. While this approach is grotesque, it is not irrational for someone who wishes to remain alive and in power at virtually any cost. Neither Ba'ath ideology nor religion had any serious role in Saddam's decision-making, although at various points in his career, he attempted to use either or both elements to justify his actions and generate support for them.

The nature of the Saddam Hussein regime was further shaped by the ways in which subordinates interacted with the leader. As noted, the persistent climate of fear dictated the information that subordinates felt comfortable providing to Saddam in ways that minimized the danger of upsetting him or appearing to contradict his personal beliefs and guidance. A near zero tolerance for criticism about him and his decision-making was a fundamental characteristic of the Iraqi regime, which often effectively prohibited serious discussion of major issues. Moreover, Saddam had by 1991 developed an inner circle of advisors who understood these unspoken rules with absolute clarity and conducted themselves in ways designed to retain the dictator's favor and not to critique his strategic thought. A further complication involved Saddam's insular life, and intolerance for alternative viewpoints allowed little opportunity for useful input from the leaders or senior statesmen from friendly Arab states or other favorably-disposed countries. He tended to view such regimes as untrustworthy or as having inferior leaders. King Hussein of Jordan was probably Saddam's most respected associate among

Arab heads of state in 1990, but he disregarded King Hussein's very good advice to him during the pre-war crisis, including a warning that the United States would probably react with force if Saddam invaded Kuwait.[131] King Hussein's deep knowledge of the West, and particularly the United States, would seem to dictate that his views be taken seriously, but Saddam felt he knew better.

The Iraqi national security decision-making approach (it is difficult to call it a system) consequently played havoc with any effort to make rational choices, but had a subtle logic for addressing Saddam's concerns about remaining in power, and especially preventing a military coup. The Iraqi dictator had made a number of huge and costly mistakes throughout his tenure in power, including some extremely bad decisions on the conduct of the 1980-88 Iran-Iraq War. If military officers were allowed to raise criticisms about Saddam's genius and his previous (sometimes disastrous) intrusions into military affairs, who knew where this could lead? Rather than allow this to happen, the Iraq government rigidly monitored its officer corps, and ruthlessly moved against those officers who appeared to offer even the most limited forms of disagreement.[132] Mild criticism of Saddam's military decisions when overheard or made over a monitored telephone network usually led to rapid imprisonment.[133] More serious criticism of Saddam himself would have more serious consequences including torture, execution, or removal of one's tongue. Failing to report criticism of the President was also a crime. Conversely, Saddam was often willing to forgive promising officers for minor acts of dissent after they had spent a limited time in prison, assuming (often correctly) that the brief prison experience followed by release, rehabilitation, and sometimes promotion would reinforce their commitment to show absolute obedience in the future.

Iraq's Initial Efforts to Assess the Danger Posed by the United States.

The Gulf crisis leading to the first U.S.-Iraqi war began on August 2, 1990, when Iraqi forces invaded Kuwait over what began as a dispute over oil and Iraq's demands for massive Kuwaiti foreign aid to help meet the economic crisis brought on by the economic problems associated with the 8-year Iran-Iraq war. By the evening of August 2, Iraqi forces had established virtually complete control over Kuwait City and were moving in force to seize the oil fields in the south of the country. Prior to the war, Saddam had carefully considered potential world reaction to such an invasion within the constraints of the Iraqi approach to policy formulation that stressed his personal judgment to an extent the dwarfed all other domestic inputs to key decisions and foreign policy initiatives.

The United States did not anticipate the Iraqi attack on Kuwait until shortly before it occurred. The leaders of friendly Arab countries, including Kuwait, requested that the United States exercise restraint in dealing with Baghdad during the crisis in the belief that Saddam was bluffing. Kuwait also declined an invitation to participate in an upcoming military exercise with the United States, an act that, once announced, would have indicated a U.S. interest in protecting Kuwait.[134] Consequently, there was no ongoing, forceful, and committed attempt to deter Iraq from seizing Kuwait by either the United States or the Arab world. Prior to the invasion, both the Saudi Arabian and Egyptian leaderships received Iraqi assurances that no invasion would occur, and such assurances appeared to reinforce their belief that diplomacy could be used to resolve Iraqi-Kuwaiti differences.[135] Egyptian President Hosni Mubarak was told that no invasion would occur in a face-to-face meeting with Saddam Hussein, where the Iraqi dictator also asked Mubarak not to inform the Kuwaitis that they were safe from such an attack. The Kuwaitis, however, were already showing a declining concern over the danger of invasion and had placed their military only on a very limited level of alert.

Saddam invaded Kuwait with little expectation of a U.S. military response. He viewed the United States as a technologically superior power but also felt that U.S. military options would be limited by a high level of casualty aversion. In one of his most well-known comments, Saddam told U.S. Ambassador to Iraq April Glaspie on July 25, 1990, "Yours is a society that cannot accept 10,000 dead in one battle."[136] In making statements such as this, Saddam drew great comfort from what he perceived to be the lessons of the Vietnam War. In particular, Saddam and other Iraqi leaders repeatedly held out Vietnam as an example of how a small and determined developing country could defeat the United States by showing resolve and inflicting significant numbers of U.S. casualties, while stoically accepting massive numbers of their own casualties.[137] He also believed that peace movements in Europe and the United States would develop into a strong domestic political force such as occurred during the Vietnam War. Saddam hoped that a strong peace movement would limit U.S. offensive options and perhaps help to end the war once the number of U.S. casualties started to escalate. Saddam further doubted that the United States was particularly concerned about the fate of Kuwait so long as it continued to have access to oil at reasonable prices. Washington did not have a defense treaty with Kuwait, and ties between the two nations were not viewed as special. At the beginning of the crisis, U.S. Assistant Secretary of State for Middle East Affairs John Kelly was asked about U.S. defense obligations to Kuwait during a congressional hearing. He responded that the United States had no formal obligations to Kuwait, despite the reflagging of Kuwaiti tankers in 1987.[138] The reflagging, he explained, was a discrete agreement that did not carry any larger implications for defending the emirate. This answer was merely a factual response to a specific question about U.S. legal obligations rather than a statement of policy. Nevertheless, Saddam heard an Arabic translation of the response within a few minutes of the statement being made, and was obviously encouraged by such a response. Kuwait, at this point in time, had reason to deeply regret the lack of formal security ties with major global powers.

While many aspects of Saddam's assessment appeared reasonable, he quickly realized that he had miscalculated the U.S. response. U.S. leadership, and especially President George H. W. Bush, reacted with fury to the Iraqi invasion of Kuwait in a way that Saddam had never expected. This reaction was not inevitable and stemmed heavily from the values and character of President Bush. While virtually all of the U.S. leadership agreed that Saudi Arabia had to be defended against any future Iraqi predations, it was not clear on August 2 that the United States was prepared to commit itself to freeing Kuwait, and that it would eventually move almost half a million troops into the region in order to do so.[139] President Bush signaled his resolution on August 5, when he stated, "This will not stand. This aggression against Kuwait."[140] Three days after this statement, the United States deployed the first

U.S. combat units to Saudi Arabia. Saddam was deeply and unpleasantly surprised when the Saudi Arabians agreed to host U.S. troops in a dramatic departure from their previous foreign policy approach which was often to seek accommodation with dangerous regional adversaries rather than confront them. As the crisis continued and the scope of Iraqi atrocities in Kuwait became known to him, President Bush became unshakable in his determination to expel the Iraqis from Kuwait.[141] The President increasingly drew analogies from the 1930s, viewing Saddam as a new Hitler who could not be addressed through appeasement.[142] Arab members of the coalition including Saudi Arabia, Egypt, and Syria, were also increasingly willing to support military action against the Iraqis due to a fear the United States might reach an accommodation with Iraq and then withdraw its forces from the Gulf, leaving Saddam with both a new set of grievances against them and an intact military machine.[143] Saddam could then use his undiminished military capabilities and enhanced prestige to threaten regional stability at some future point in time.

As the potential for war with the United States increased, Saddam became deeply involved in planning for a military conflict. He showed a great deal of confidence in his own strategic vision and did not seem to have any reservations about intervening in the conduct of military operations. General Hussein Rashid Mohammad, the Iraqi Armed Forces Chief of Staff, stated that Saddam's guidance was "continuous" and that the military staffs followed this guidance "to the tiniest detail."[144] This highly centralized approach whereby the Iraqi military is completely subordinate to the whims of an amateur strategist was clearly the model that Saddam was most comfortable in implementing. Conversely, the Iraqi dictator tended to view any general who distinguished himself or became a war hero as a dangerous potential rival to the president, and such individuals were sometimes arrested or more often required to retire. An additional problem complicating Saddam's poor understanding of modern warfare was the lack of a steady stream of reliable intelligence analyzed and presented by intelligence professionals. Even during the war itself, with the regime's existence potentially at stake, frankness in dealing with Saddam could still be fatal. This situation meant that he was making key wartime decisions on the basis of information that had been deliberately distorted to make Iraqi military prospects to appear more hopeful than they actually were. It is, for example, unclear that Saddam fully comprehended the heavy damage inflicted by the allied air campaign until it was brought to his attention by Yevgeni Primakov, a Soviet diplomat seeking to mediate the crisis at the request of President Mikhail Gorbachev.[145]

Saddam also appeared to believe that the perceived U.S. aversion to casualties would limit the scope of the conflict to an air campaign if war actually broke out. He understood that the United States possessed significantly better aircraft technology than the Iraqis, but he also may have expected more from Iraq's air defense forces. Despite these expectations, coalition losses to Iraq's SAMs were extremely rare due to the effectiveness of electronic countermeasures directed against them. By the end of the war, only around 10 coalition aircraft had been shot down by SAMs despite thousands of these missiles being fired.[146] Saddam also did not seem to take superior U.S. training for ground forces into account. He made statements suggesting that Iraqi soldiers were better than U.S. soldiers due to their battlefield experience in the Iran-Iraq war. Living with his delusions throughout the war, he later claimed that the Battle of Khafji (which Saddam surprisingly claimed and appears to have viewed as a victory) reinforced this understanding, and that the United States was unprepared to risk significant casualties in combat with such adversaries.[147] While Saddam had many delusions about the U.S. military, he understood that the Iraqis were badly overmatched in the field of airpower. This knowledge was rapidly reinforced in the opening phase of the war. In contrast, he suspected that the United States would not wish to commit large numbers of troops to ground combat, especially urban combat where casualties can be significant.[148]

Iraqi Chemical, Biological, Missile, and Nuclear Capabilities in 1990.

At the beginning of the Kuwait crisis, Iraq was known to possess weapons with some potential to deter an attack by a powerful adversary. By 1991 the world had been provided with a considerable opportunity to observe the growth and development of Iraqi chemical warfare (CW) capabilities. As far back as the early 1980s, the Iraqis began using chemical weapons during their war with Iran.[149] They also used chemical weapons against their own Kurdish citizens in Halabja in 1988 where local Kurds were suspected of cooperating with Tehran. Chemical weapons are sometimes viewed as the vital factor that allowed the Iraqis to prevail in a series of military engagements toward the end of the war. By this time, the Iraqis had fully and effectively integrated chemical weapons into their offensive and defensive military doctrine.[150] According to one analyst, the Iraqi use of chemicals was such that it allowed them to "gas the war to an end."[151]

Iraq obtained unmodified *Scuds* in huge quantities from the Soviet Union during the Iran-Iraq war and used these systems extensively in combat.[152] In the final 5 months of the war, they had also bombarded Iranian cities with extended-range *Scud* missiles in a conflict known as the "war of the cities." While hundreds of these systems were fired at Iran in the war, all were equipped with conventional warheads. Chemical weapons had been used by the Iraqis since 1983, but they were not used in conjunction with these missiles which could strike targets over 600 kilometers from the launch site. An ongoing effort to develop chemical warheads would probably have led to the use of these warheads in combat if a cease-fire had not been put into place in August 1988. Such warheads could have increased the value of the missiles since *Scuds* have relatively small payloads, and a highly toxic warhead could be more effective than conventional explosives for producing mass casualties. Furthermore, extended range *Scuds* had an even smaller payload than unmodified *Scuds*. Interestingly, the use of hundreds of *Scuds* against Iran during the war of the cities led to only around 2,000 deaths and 6,000 wounded, although they functioned as a highly effective terror weapon.[153]

The Iraqis made extensive use of chemical weapons on the battlefield during the Iran-Iraq war but are not known to have used biological weapons in combat. This situation made it more difficult to ascertain what Iraqi BW capabilities actually were. The Baghdad leadership did not claim to have weaponized biological agents, although Saddam maintained that it was Iraq's right to develop any weapons that Israel possessed (presumably he included biological weapons).[154] Instead, in the lead up to the war, the Iraqis did make numerous claims of secret weapons and "surprises" which would be revealed during the course of the fighting. While the Iraqis did not identify biological weapons as the subject of these references, this seemed a particularly worrisome possibility. In his memoirs, General Colin Powell refers to "CIA [Central Intelligence Agency] estimates" that presented a "troubling" picture of Iraq's potential for biological weapons use which he contrasted with what he called the "manageable" threat of chemical weapons.[155] Additionally, a number of Western news reports and specialty defense-related publications consistently indicated that Iraq had been pursuing a BW capability even prior to the conclusion of the Iran-Iraq war in August 1988.[156] The Iraqis may have believed that the United States would assume that Baghdad possessed a biological weapons capability and chose to be vague about this capability because they felt this approach would work more effectively for deterrence than explicit threats.

After the 1991 war, the outlines of Iraq's BW capabilities became clear. A strong BW research program was initiated during the Iran-Iraq war, and the Iraqis were clearly contemplating an escalation to BW use should the war with Iran threaten the survival of Saddam regime. After the cease-fire with Iran, this research was continued and expanded.[157] The biological agents that Iraqi scientists found the most promising were anthrax and botulinum toxin which they later produced in bulk. Anthrax is one of the hardiest biological weapons known to science. Botulinum toxin is drawn from a living organism and is therefore classified as a biological weapon despite the fact that it is not a bacteria or virus. In many respects, toxins are more akin to chemical weapons than they are to other forms of biological weapons, but botulinum toxin is staggeringly more toxic than any known chemical weapon. This situation suggests that Iraqi expertise in utilizing chemical weapons could be of some value in using botulinum toxin. In the months leading up to the 1991 war, the Iraqi BW program produced 8,445 liters of *Bacillus anthracis* which causes anthrax and 19,180 liters of *Clostridium botulinum* toxin. They also produced aflatoxin, a long-term carcinogen which would be inconsequential on the battlefield. It is not clear why the Iraqis viewed the third agent as promising.

In April 1995, well after the war had ended, the Iraqis finally admitted to UN inspectors that they had 166 gravity bombs and 25 missile warheads configured to be armed with biological agents.[158] This admission was made in response to the defection of Saddam's two sons-in-law, Hussein Kamil al Majid, and his brother, Saddam Kamil al-Majid, to Jordan. Hussein Kamil had previously headed the Military Industrial Commission (MIC) which included the biological warfare program. The willingness of these defectors to speak openly with Western intelligence officers and the world press rendered it impossible for the Iraqi government to continue its policy of denying the existence of a biological warfare program.[159] Hussein Kamil revealed the existence of extensive numbers of documents relating to the Iraqi BW program at his chicken farm in Iraq. The UN Special Commission (UNSCOM) which helped to monitor the WMD and missile disarmament of Iraq was able to obtain these documents in 1995.[160] Incredibly, in February 1996, Hussein Kamil and his brother were persuaded to return to Iraq where the regime quickly had them assassinated.

Saddam also maintained a nuclear weapons development program prior to the Kuwait crisis, but he was at least a year away from an operational nuclear weapon in 1990. When the crisis over the invasion of

Kuwait broke out in August 1990, Saddam ordered a crash program to build a nuclear weapon, but it was not possible to accelerate the schedule so that this kind of device could be finished in time to be operational during the conflict and thereby serve as a deterrent or a war-fighting weapon once the coalition offensive began in January 1991.[161] Saddam's only possible option in the nuclear field was to assemble some sort of radiological weapon (a "dirty bomb") of doubtful effectiveness. Such a weapon would be of uncertain military value and almost certainly fail to improve Iraqi chances for fighting the United States to a standstill. Conversely, the use of a weapon such as this (with its special negative stigma) would have tremendous potential to incite a harsh coalition response.

It is uncertain that Saddam understood the shortcomings of his WMD systems given the predilection for dishonesty by his subordinates, but he could not have been ignorant of all of Iraq's potential technological problems. He had, after all, demanded that Iraqi scientists not only build a nuclear weapon (which they could not do in the time frame he required), but also to show dramatic improvement in the weaponization of biological agents. These demands, however unreasonable, indicated an awareness of serious shortcomings. Saddam was further aware that the United States had sophisticated chemical/biological defensive equipment that would reduce troop vulnerability to Iraq chemical weapons. During the Iran-Iraq war, even the Iranians had been able to reduce greatly their chemical casualties with much more rudimentary defenses.[162] Thus, he had reasons to doubt the military effectiveness of his chemical and perhaps untested biological agents when employed against protected troops. This level of knowledge may have helped Saddam resist any inclination to use chemical weapons.

Iraq also faced a number of potential difficulties in using long-range systems for delivering chemical or biological weapons for strategic rather than battlefield use. The Soviet Union had rejected Iraqi requests to purchase *Scud* missile warheads suitable for the delivery of chemical or biological weapons. In response, Baghdad sought to develop indigenous warheads to be used with such weapons. After the war, UN arms inspectors described these warheads as primitive and of questionable value. Some of the inspectors stated that the warheads were so poorly designed and were unevenly filled with chemicals that they could not be fired accurately, if at all.[163] Since these warheads were not used in the 1991 combat, it is uncertain how aerodynamic they were and how well they could be targeted if they were fired at Israeli, Saudi Arabian, or U.S. military targets from Iraq. Moreover, *Scud* missiles can only deliver a limited payload, and it is doubtful that these unsophisticated warheads could have had much impact beyond serving as a terror weapon since they may not have been able to disperse deadly microbes, chemicals, or toxins. It was also uncertain if Iraq's warheads were designed effectively enough for living organisms to survive rather than simply be burned off when the weapon explodes. More sophisticated chemical warheads have an airburst capacity allowing them to disseminate chemical agents widely in order to inflict as many casualties as possible. The Iraqis did not have this capability. Additionally, there were also questions about the R-400 gravity bomb which the Iraqis designed to carry botulinum toxin. According to a number of UN inspectors who examined the system, it was configured in such a way that almost all of its contents would be destroyed by pressure and heat if it was used in combat.[164]

Iraq, therefore, had some WMD capabilities as it was facing the prospect of war with the United States, but the weapons that it possessed were problematic for both deterrence and warfighting. Saddam had to make the most of these capabilities in his effort to deter the United States, but many of the limitations of his arsenal were apparent and difficult to ignore. The weapons that most clearly worried U.S. planners were the undefined biological agents. These systems clearly had the greatest deterrent potential, but there were drawbacks for Baghdad in declaring these weapons. Saddam was in the awkward position of threatening the U.S.-led coalition with products of a weapons program that threatened to expand and become more dangerous over time unless external forces prevented this development. Eliminating Saddam's WMD, and especially his nuclear and biological weapons programs, thus became a key allied war goal. The more he blustered about his missiles and unconventional weapons, the more important this goal became to U.S. and coalition leaders. Aspects of this logic became relevant again in 2003 when the United States waged a new war against Iraq to eliminate WMD, which by that time no longer existed.[165]

Iraqi Efforts to Deter a U.S. and Coalition Attack Without Withdrawing From Kuwait.

In addition to Saddam's behavior during the Iran-Iraq war, the background for how U.S. leaders thought about Iraqi WMD capabilities was also set by a speech that he gave on April 2, 1990. On this occasion, the Iraqi dictator appeared to threaten to use chemical or biological weapons against Israel in his lecture to the General Command of the Iraqi Armed Forces. His language on this occasion was vitriolic but hazy. He threatened to "burn" half of Israel if the Israelis attacked Iraq. Burn is not a term that is usually used with the most advanced forms of chemical weapons or with biological weapons. The reason for choosing this occasion to give such a speech is unclear, but Saddam may have felt that the Israeli intelligence service had uncovered information about his ongoing effort to produce weapons grade uranium for use in primitive nuclear weapons.[166] If so, Saddam's concern about a bombing attack may have been realistic and based on former Israeli Prime Minister Menachem Begin's 1981 pledge to use military means to prevent the acquisition of nuclear weapons by Israel's regional enemies ("Begin Doctrine") made in the aftermath of the destruction of Iraq's Osirak nuclear reactor by the IAF. At the time, Saddam had been acutely embarrassed by Iraq's inability to respond to the attack in any serious way. Underscoring this humiliation was his order to execute the commander of the air defense units around the facility for his negligence in failing to protect the reactor.[167]

The Israeli leadership responded to Saddam's April 2 speech with a number of statements suggesting that any attacker would be subjected to overwhelming and disproportionate responses by Israeli forces.[168] While the word nuclear was not used, the meaning of the Israeli threats appeared unmistakable. Some Israeli scholars have also suggested that it is possible that Israel sent secret communications to the Iraqi leadership promising massive nuclear retaliation in ways that were impossible to misinterpret.[169] Publicly Yitzhak Rabin, a Labour Party leader, who had just left his position as defense minister said, "We have the means for a devastating response, many times greater than Saddam Hussein's threats."[170] This statement was actually milder than some of his earlier promises about an Iraqi chemical weapons attack. In July 1988, Rabin stated that the response to an Iraqi attack would be to "hit them back 100 times harder."[171] Such threats might have been a useful background experience for Saddam to consider later in that same year when he was in the awkward position of trying to provoke Israel into war without triggering their willingness to use nuclear weapons. Moreover, by this time, Israel had experienced something of a nuclear weapons scandal when Mordechai Vanunu, a former technician at the Dimona nuclear facility, went public in 1986 with his description of the Israeli nuclear weapons program including a variety of pictures, which he sold to *The Sunday Times*. Vanunu was later abducted by Israeli intelligence agents, convicted of treason, and imprisoned for 18 years.[172] This scandal could not have failed to make an impression on Saddam.

Many of Saddam's 1990-91 efforts to discourage the United States from attacking Iraq and Iraqi forces in Kuwait did not involve explicit public threats of chemical or biological weapons use under any wartime circumstances. While Iraqi leaders made threats about chemical weapons use, such warnings were often (although not always) conditioned on the U.S. refraining from nuclear weapons use rather than refraining from initiating a war. As such, threats of chemical weapon use did not constitute the core of Saddam's strategy for deterring war with the United States and its allies. Rather, he hoped to convince the U.S. leadership that any war fought on the conventional level would involve massive numbers of casualties, and that the United States might want to consider seriously the cost of such a struggle before embarking upon it. Saddam understood that Iraq could only inflict a small fraction of the casualties on the United States that could be inflicted upon Iraq, but he also firmly believed that the Iraqi military could accept a great deal of battering while the United States had a very low threshold for pain. According to captured documents, Saddam told an assembled group of officers that the United States and its allies did "not have the same level of determination as the Iranian enemy."[173] He further stated and may have believed that U.S. forces would "collapse" if confronted "in a determined way."[174] Saddam also told his staff that Iraq would achieve victory if it obtained "a ratio of four Iraqi casualties to every one American."[175]

Thus, Saddam's flawed analysis and faulty decision-making process caused him to believe that he had a rational and achievable strategy for victory using only conventional weapons. A more informed analysis would have cast doubt on both the concept of a limited U.S. threshold for casualties (in a popular or at least not unpopular war) and the ability of Iraq's conventional forces to inflict mass casualties on the United States in any potential conflict. Such analysis would have challenged his core belief that a war against the United States would assume many of the same characteristics of the Iran-Iraq war. During this conflict, there were horrendous levels of casualties on both sides and long periods of stalemate and trench warfare. The idea that the United States would fight Iraq in the same way as Iran had waged war seems quite fanciful.

Beyond their effort to deter or shorten the war with the threat of massive casualties in conventional fighting, the Iraqis, as noted, also sought to emphasize the WMD threat that the United States faced if

Washington escalated the war to include nuclear weapons or presumably if U.S. ground forces attacked Baghdad and threatened to remove the Iraqi regime. The first authoritative Iraqi statement about chemical weapons use in any upcoming confrontation with the United States was issued by Foreign Minister Tariq Aziz on August 18, 1990. Aziz stated that in any war with the United States, Iraq would not use its chemical weapons unless the United States used nuclear weapons first.[176] Later, on December 22, 1990, Sa'di Mahdi Salih, the Speaker of the Iraqi National Assembly, made a statement that Iraqi would use chemical weapons if it was attacked by the UN coalition.[177] He made no mention of nuclear weapons as the trip wire for such a response and therefore seemed to be lowering the threshold of response previously established by Aziz. This contradiction might have been an uncoordinated effort or there might have been some effort to muddy the waters. Aziz, nevertheless, was a more important spokesperson, and Salih's chief responsibility during the crisis was to appear outraged about U.S. policies and display enthusiasm for Saddam's leadership. After the war began, Saddam, as noted below, reiterated the Aziz formula of chemical weapons use only in response to U.S. nuclear weapons use.

To make his WMD threats appear more credible, Saddam needed to appear able to use chemical or biological weapons in a strategic role. This meant having the capability to deliver WMD beyond the front lines with either aircraft or missiles. Unfortunately for him, the Iraqi Air Force was not much of a deterrent or a particularly good choice for delivering unconventional munitions. During the Iran-Iraq war, the air force had a number of problems, and it had not improved substantially as the result of the combat experience gained in that conflict. In the initial years of the Iran-Iraq war, Iraqi aircraft were consistently outperformed and shot down by Iran's U.S.-made fighters and especially F-4 *Phantoms*. Later, in response to dwindling Iranian stocks of aircraft spare parts and no U.S. replacement aircraft, the Iraqis achieved air superiority by default.[178] Many of Iraq's aircraft were modern, but air force leaders were ordered to use tactics designed to preserve aircraft. Most Iraqi pilots consequently gained only limited useful experience in the conflict. Often during the Iran-Iraq war, Saddam appeared more interested in preserving the air force as a force in being than in utilizing it in the most effective way he could to wage war. The Iraqi Air Force also underperformed during the initial invasion of Kuwait when some Iraqi armored and infantry forces, including Republican Guard units, were bombed and strafed by Kuwaiti aircraft without challenge from their Iraqi counterparts. Ground forces commanders were livid at the air force for being unable to protect the advancing units.[179] More importantly, the Iraqi Air Force failed to bomb Kuwaiti civilian airports to the point that members of the ruling family were unable to escape from the country by civilian aircraft, while others departed by car caravan to Saudi Arabia.[180] The ability of most of the Kuwaiti leadership to reach Saudi Arabia, where they helped to organize resistance to the Iraq annexation of Kuwait, became a serious problem for the Iraqis later during the period leading up to war in January 1991.

The shortcomings of the Iraqi Air Force therefore placed the burden of both pre-war deterrence and possible later strategic strikes squarely on shoulders of Iraq's missile force. During the pre-war period, Saddam strongly emphasized the strength of his missile units in a variety of ways, including threatening to use missiles against Israel and against any country that accepted Western troops or supported the U.S.-led coalition.[181] The threat to use *Scuds* against Israel was unrelated to any potential Israeli involvement in the war against Iraq. Rather, the Iraqis sought to widen the war in a way they hoped would provoke popular unrest in Arab countries supporting the United States. In late 1990, Saddam (truthfully) told Spanish television that Israel would be one of the first targets for his missiles.[182] When asked in Geneva shortly before the war if Israel would be attacked during hostilities, Iraqi Foreign Minister Tareq Aziz replied emphatically, "Yes. Absolutely yes."[183] Threats to use missiles against Arab states accepting Western troops were designed to intimidate Saudi Arabia and also to frighten some of the smaller Gulf Arab states, including Bahrain and Qatar, which were providing important logistical assets to the coalition.

Beyond public explicit and implicit threats of missile use, Saddam also attempted to signal his offensive power through highly visible missile tests. Prior to the war, *Scud* missiles without functioning warheads were test-fired on three separate occasions on Iraqi territory, using Iraqi airspace.[184] Since hundreds of these missiles had been utilized during the last phase of the Iran-Iraq war, one can safely assume that these launches were occurring for reasons other than the technological improvement of existing missile stocks. Another provocative signal was that the missiles were tested in the general direction of Israel rather than Saudi Arabia. Throughout the crisis leading up to the war, Saddam talked about missiles a great deal. In response to an Israeli-Palestinian confrontation and riot on Jerusalem's Temple Mount (or al-Haram al-Sharif) on October 8, 1990, an Iraqi spokesman read Saddam's response to the incident in which 19 Palestinians died. According to the announcer, the Iraqis had developed a new long range missile to deal with the Israelis and could "reach the targets of evil."[185] What is unusual about this statement is that it was already widely known that Saddam had extended range *Scuds* that could hit Israel. While he now claimed to possess a new system (which he called the *hijara* or stone, after the stones thrown by Palestinian youths in the uprising known as the Intifada), he did not claim it had any different characteristics from the extended range *Scuds*. Saddam simply wanted to use the Palestinian deaths as an opportunity to reiterate the strength of his missile force and to emphasize Iraq's solidarity with the Palestinians.

Efforts to Deter Iraq from Using Weapons of Mass Destruction Before and During the Conflict.

During the time frame when Saddam was attempting to deter a U.S.-initiated war, the United States was in turn seeking to deter Iraqi use of chemical or biological weapons should war break out. In the time frame between Iraq's August 2, 1990, seizure of Kuwait and the onset of Operation DESERT STORM on January 17, 1991, there was considerable communication, posturing, and signaling between the U.S.-led coalition and Iraq. Much of the Iraqi posturing activity has already been noted. The United States, however, was also quick to assert its concerns and thereby draw "red lines" which Saddam could cross only at the gravest risk to his regime. On August 8, 1990, the first units of U.S. ground troops arrived in Iraq as part of what was then known as Operation DESERT SHIELD. One day after these troops arrived, President Bush stated that Iraqi use of chemical weapons against U.S. military forces would be "intolerable," and that it would be dealt with "very, very severely."[186]

Other nations involved with the crisis were also concerned about Saddam's potential use of chemical weapons and added their weight to the U.S. warnings about such courses of action. In an important allied initiative, the French and British governments issued a November 1990 joint statement at the UN warning Iraq against "initiating the use of chemical or biological warfare."[187] This statement was followed by a formal pronouncement from Soviet Foreign Minister Eduard Shevardnadze warning the Iraqis not to engage in chemical warfare.[188] Although the Soviet Union was not part of the military coalition arrayed against Saddam, it was an important power which was clearly lining up with the United States over its former Iraqi ally. Arab allies of the United States were also quick to assert the dire consequences of Iraqi chemical weapons use. Saudi Arabia, in particular, issued threats about the danger of an Iraqi chemical weapons attack, although the consequences were usually left ambiguous.[189]

One of the last and most important pre-war warnings about Iraqi chemical weapons use before the war occurred at the January 9, 1991, meeting between Secretary of State James Baker and Iraqi Foreign Minister Aziz when Secretary Baker presented Aziz with a letter to Saddam Hussein from President Bush. The letter (which was subsequently provided to the media) contained the following statement:

> Let me state too, that the United States will not tolerate the use of chemical or biological weapons, support of any kind of terrorist actions, or the destruction of Kuwait's oil fields and installations. The American people would demand the strongest possible response. You and your country will pay a terrible price if you order unconscionable actions of this sort.[190]

Baker also emphasized this point in his discussions with Aziz, "If there is any use of weapons like that, our objective won't just be the liberation of Kuwait, but the elimination of the current Iraq regime and anyone responsible for using those weapons would be held accountable."[191] While the United States also made it clear that it considered the destruction of Kuwaiti oil infrastructure and acts of terrorism abroad unacceptable, the real emphasis was on CW deterrence which Baker addressed in what he called "the bluntest possible terms."[192]

In his discussions with Aziz, Baker attempted to elevate the threat to respond to Saddam's use of chemical weapons beyond the level of simply a signal or a threat. He sought to stake U.S. credibility on this statement by framing his assertions about future courses of actions as a "pledge." The implication of this wording is more than semantics. By placing U.S. credibility on the line so unequivocally, the U.S. Secretary of State was indicating the United States could not fail to act on its promise to retaliate harshly to chemical weapons attack without having its credibility damaged in future diplomatic encounters with a variety of nations over an assortment of topics. Baker was deliberately locking the United States in to a specific course of action in a way that he hoped would influence Saddam to show restraint. Moreover, the United States did not offer even the shadow of a compromise over any key differences with Iraq. As a diplomatic exercise, the U.S. sole focus at Geneva was to demand Iraqi acceptance of all relevant UN Resolutions and to make the Iraqis aware of red lines that the United States was drawing. Also, while the focus of this discussion was chemical weapons, it can be safely assumed that the United States would react just as ferociously to the use of biological weapons. Since the Iraqis had not yet openly admitted that they possessed biological weapons, the threat was made by implication, but all of the reasons that chemical weapons were deemed unacceptable applied at least equally to biological weapons.

The war between Iraq and the U.S.-led coalition began on January 16, 1991 around a week after the Baker-Aziz meeting in Geneva. The war was initiated with numerous U.S. air strikes and cruise missile attacks against large numbers of Iraqi regime and military facilities. Saddam responded with conventional missile strikes against coalition and Israeli targets. The beginning of the air war did not put an end to an effort to deter Saddam's use of chemical or biological weapons. On February 1, 1991, Vice President Dan Quayle told BBC television that the U.S. policy was "very clear" and "we simply don't rule any options in or out."[193] He later sent somewhat more mixed signals by suggesting that he could not imagine President Bush using nuclear or chemical weapons but continued to assert that all options were open. In response to these apparent contradictions, Secretary of Defense Richard Cheney was asked to clarify the policy that the Vice President was outlining. Cheney sarcastically responded, "I think it means we don't rule options in or out."[194] In this interview and interviews that followed, Cheney especially reiterated not ruling out nuclear weapons use.

President Bush entered the fray again on February 5 as the air war raged but before the ground assault had been initiated. The President stated that "he [Saddam Hussein] ought to think very, very carefully" when considering the use of chemical weapons.[195] President Bush would not further explain this statement, saying that he preferred to leave Hussein with a "very fuzzy interpretation" of what his options might be. A few days later Secretary of State Baker, elaborating on this theme during congressional testimony, stated that "The use of [chemical] weapons would have the severest possible consequences."[196] This renewal of strong U.S. rhetoric occurred after several serious events that may have influenced the Bush administration. The first of these was a Cable News Network interview of Saddam Hussein by Peter Arnett in which the Iraqi dictator stated, "I pray to God I will not be forced to use [chemical] weapons, but I will not hesitate to do so should the need arises." He did add in his own intrawar deterrent effort, "We shall use the weapons that will be equitable to weapons used against us by our enemies."[197] Here, Saddam seemed to be echoing the earlier Aziz statement that chemical weapons would only be used in response to nuclear strikes. Still, any assertion about "not hesitating" to use chemical agents had to be taken seriously. The second event of concern was the capture of a number of Iraqi soldiers at the January 29-February 1 Battle of Khafji, where 304 out of 429 prisoners were carrying gas masks.[198]

U.S. public opinion polls during the war and the lead up to it showed a large number of Americans willing to support or accept the use of nuclear weapons if such usage saved significant numbers of American lives. One political group even took out a number of full page ads in national newspapers advocating the use of nuclear weapons against Iraq even if Baghdad did not use chemical or biological weapons against U.S. forces.[199] Newspaper reports of public views on this issue and even the ads may have had a disproportionate influence on Iraqi thinking since the Iraqi intelligence organizations relied heavily on the U.S. press and mass media for information on U.S. capabilities and potential courses of action.[200] By this time, with the air war raging, Saddam may have realized that his hopes for the development of a powerful and constraining U.S. peace movement had come to very little. U.S. domestic politics did not look like a strong bet to restrain nuclear weapons use.

Saddam probably believed that the United States would use nuclear weapons in some scenarios for a variety of reasons and would have been attentive to the statements and signals mentioned above. Certainly Saddam would never have viewed moral restraint and issues of a disproportionate use of force as serious reasons for Washington to forego such a decisive military advantage. While the United States might pretend to agonize or equivocate over this option, in the end they would continue to view nuclear weapons as a serious and useful option. Moreover, the United States had previously used nuclear weapons against Japan in 1945 on two separate occasions. While these nuclear strikes were a chronologically distant event, they were a precedent that was difficult to dismiss entirely. Perhaps with these circumstances in mind, Saddam made the decision to have the Iraqi media provide the public with elaborate instructions on protective measures against nuclear and chemical attack.[201] It is doubtful that Saddam would have alarmed and perhaps demoralized Iraq's population with something as alarming as nuclear civil defense instructions if he did not believe that the threat was serious.

In addition to the dynamics between the United States and Iraq, there was also the issue of Israel. The initiation of U.S. and coalition air strikes against Iraq led to almost immediate Iraqi extended-range *Scud* missile strikes against Israel in accordance with standing orders from Saddam.[202] U.S. sources state that Iraq fired 39 missiles against Israel throughout the war, while the Iraqis claimed to have launched 43.[203] Saddam predelegated authority to his missile force commanders, which required them to begin firing conventional missiles at Israel as soon as war broke out. These commanders were ordered to fire large numbers of missiles at Israel, but they were forbidden to use chemical or biological warheads unless explicitly authorized to do so by Saddam himself. His purpose in attacking the Israelis was to provoke a military response from them that would widen the war to include Israel and thereby threaten to undermine Arab cooperation with the United States in the liberation of Kuwait. He also hoped that various Arab governments might be overthrown should they continue to cooperate in a war in which they fought in a de facto alliance with Israel.

Israeli security policy called for an immediate—almost automatic—retaliatory strike against any Arab state attacking Israeli population centers. Such a response appeared especially likely in this instance after an Israeli civil defense team using chemical detection equipment reported that they had identified traces of nerve agent within the missile. It therefore appeared possible that the first Iraqi missile strikes against Israeli population centers included a missile that was carrying a chemical warhead but had failed to produce casualties because of a malfunction or some other problem. Later, before a strike was ordered, it was discovered that the team had taken a false reading from the fuel tank of one of the *Scuds*.[204] This development allowed U.S. leaders to press their case for Israeli restraint to avoid playing into Saddam's hands, while the U.S. military dealt with Iraq. No chemical agent had been used in the attack.

Iraq fired modified longer-range *Scuds* at Israel throughout the war, and it is deeply uncertain that one

Israeli reprisal strike would suffice to punish Iraq once the cycle of attack and respond had begun. Moreover, there was also the question of how Israeli aircraft were going to reach Iraq. According to Efraim Halevy, the former Director of Israel's Mossad intelligence service, secret meetings with Jordan's King Hussein quickly led to a Jordanian promise that Amman would not allow Iraqi aircraft to overfly Jordan on their way to attack Israel.[205] The king would not, however, accede to an Israeli request that the IAF be allowed to use Jordanian airspace as a route to Iraq to respond to *Scud* attacks. He stated that he could not allow himself to be seen as colluding with Israel against an Arab state.[206] This development complicated the nature of retaliatory strikes since the IAF might have to brush aside Jordanian resistance when overflying Jordanian territory. While the Jordanian Air Force would not present a substantial challenge to the Israelis, the political complications of shooting their way through Jordan would further complicate the efforts by President Bush and Secretary of State Baker to hold the international coalition together.

The Israelis also had to consider the possibility that an escalating cycle of violence with the Iraqis could lead to Baghdad's use of chemical or biological weapons against them, although this possibility was remote. It would probably only occur if Iraq was able to fire some remaining missiles after Israel initiated the kind of devastating nuclear strikes which they publicly hinted about in April 1990. While Saddam wanted to provoke a serious conventional Israeli response, he did not want to provoke an apocalyptic one. Firing conventional missiles into Israeli population centers was expected to produce a disproportionate Israeli response, but it would be difficult to justify a nuclear response without large scale casualties or the use of chemical or biological weapons. Use of chemical weapons, on the other hand, seemed the most likely way to provoke the Israelis into a nuclear strike even if those weapons failed to produce mass casualties. If they did produce mass casualties, escalation to a nuclear level may well have been unavoidable. When the first *Scud* attack occurred against Israel, U.S. Deputy Secretary of State Lawrence Eagleburger (who was frequently in Israel throughout the conflict coordinating with the Israeli government) stated, "I know these people [the Israelis]. They're going to retaliate. If it's nerve gas, we'll never stop them."[207] While Eagleburger did not claim Israel would use nuclear weapons in retaliation, the nerve agent scenario he outlines makes it seem likely. Moreover, a weak and timid response to an Iraqi nerve agent attack would be viewed by the Israelis as a ghastly precedent for future deterrent efforts—almost worse than no attack at all.[208]

Saddam was also receiving significant amounts of misinformation about large numbers of coalition casualties by subordinates who were afraid to tell him the truth or may have misunderstood the situation themselves.[209] Inflicting mass casualties without using unconventional weapons was exactly the strategy that Saddam sought to utilize to serve his larger goal of increasing U.S. losses to the point that the United States would seek a negotiated end to the conflict. Nevertheless, large numbers of U.S. casualties were not an end to themselves; they were a way to help undermine U.S. national will, which Saddam had identified as the key U.S. vulnerability. Using chemical weapons may have inflamed rather than undermined U.S. public opinion and willingness to continue with the war. Saddam had previously faced considerable international criticism for using chemical weapons against the Iranians and against the Iraqi Kurds. While such criticism was mostly an inconvenience, it may have alerted the Iraqis to the potential for international sensitivity over these weapons.

There were also military effectiveness issues. The battlefield use of CW would not have led to Iraqi tactical victories, but it could have conceivably led to increased U.S. casualties, which was exactly the outcome that Hussein hoped to use to press the United States to discontinue the war. Nevertheless, he held back. In addition to the fear of retaliation, he might also have realized that there were a number of battlefield constraints that may have reduced the effectiveness of Iraqi chemical weapons use.[210] As previously noted, Saddam's experience during the Iran-Iraq War was particularly compelling in underscoring the shortcomings of chemical weapons in battle. Underlying much of his decision-making was the belief that he could perhaps retain power if it the war remained conventional simply because he was willing to accept large numbers of casualties while he believed that the United States could not do so. Saddam's uncertainty about the value of chemical and perhaps untested biological weapons, coalition threats of retaliation, and his strong belief that he had a path to victory using conventional warfare thus continued to dominate his approach to the ongoing fighting.

Iraqi Efforts to Deter U.S.-Led Regime Change.

While Saddam could not prevent the United States from liberating Kuwait and capturing southern Iraq, he clearly hoped that he could deter it from seeking to seize Baghdad and then removing his dictatorship by force. The Iraqis later characterized the survival of Saddam and his regime as a victory attributed to deterrence of the United States due to a U.S. fear of mass casualties (rather than a reluctance to seize, occupy, and administer Iraq). As previously noted, Saddam singled out the Battle of Khafji as a turning point in the war where supposedly massive numbers of coalition soldiers were killed, causing the U.S. leadership to reconsider moving forward to Baghdad. This evaluation was extremely faulty since Khafji involved very few U.S. casualties, including 25 American dead, 11 of whom were killed by friendly fire.[211] Saddam, however, had been led to believe that the United States had been severely punished in this battle and would therefore be forced to reconsider moving on to Baghdad because of the expected cost in casualties. At this point in time, Iraqi troops also had considerable recent experience in urban warfare against the Iranians, and there was at least a chance that they would put up a reasonable fight against advancing coalition forces once the fighting moved into the cities.

On January 28, Saddam also threatened to use Iraq's chemical weapons, and perhaps by implication its biological weapons, in response to the use of unspecified "equitable" weapons by the coalition by which one may assume he meant tactical nuclear weapons. He also claimed that he would not use them if the United States did not use its unconventional weapons. These statements to a reporter with a worldwide audience, including leaders of the United States, seem to indicate that the Iraqi leader had his own approach to intrawar deterrence and was signaling this approach to his adversaries. Saddam, like President Bush, wanted to keep the war within bounds. While some of his comments to the Soviets indicate that he probably understood that Iraq would lose Kuwait, he still almost certainly believed that his regime had a good chance of surviving the conflict.[212] Therefore, Iraqi restraint was in order in accordance with Secretary Baker's warning in Geneva. Baghdad, however, did renew the effort to redraw its own red lines. In mid-February, Iraq's UN Ambassador Abdul Amir al-Anbari renewed threats of undefined WMD use in response to "high-altitude bombings."[213] This statement was made as it became increasingly clear that Iraqi military forces has been severely defeated, and the regime was now vulnerable. Saddam needed to signal that he was still willing to escalate the conflict if the coalition moved to seize the capital.

The United States unilaterally declared a preliminary cease-fire with Iraq on February 28, 1991. Iraq was defeated and expelled from Kuwait following a 42-day war, which included a ground campaign of a little over 4 days. The war ended with Iraq's acceptance of UN Security Council Resolution 687. President Bush and National Security Advisor Brent Scowcroft believed that there was a good possibility that the magnitude of Saddam's defeat would help to push the Iraqi armed forces to overthrow him.[214] Massive military outrage over the defeat and the unraveling of the cult of personality seemed a serious possibility in the aftermath of the cease-fire as Saddam was exposed as a bungler who made mistake after mistake during the high stakes confrontation. Looking ahead, Iraqis might consider that Saddam had decided on a disastrous war, and his continued presence could only undermine Iraqi attempts to rejoin the world community as something other than a pariah state. To underscore this point, President Bush called upon the Iraqi people and the Iraqi Army to rise against the dictator in what he hoped would result in a military coup. Since Saddam brooked no opposition or even serious questioning of his policies, any new Iraqi government could make a case that Saddam and his immediate entourage were solely responsible for the invasion of Kuwait. A post-Saddam military government could claim that his arrest, or more probably his death, should relieve them from the burden of economic sanctions and other grinding post-war punishments. Correspondingly, the opportunity to end Saddam's rule due to his egregious lack of judgment was combined with the implicit offer of a much better post-war Iraqi future if the army did move on this opportunity. A unique opportunity seemed in the offing.

It is now well known that the U.S. administration's seemingly elegant and logical strategy for ousting Saddam without seizing control of the country was unsuccessful due to a U.S. failure to comprehend fully the fragmented nature of Iraqi society. In the aftermath of the defeat in Kuwait, popular uprisings took place in the Kurdish north of Iraq and the Shi'ite south. President Bush's statements clearly encouraged these uprisings, although it is unclear that absence of such rhetoric would have prevented the uprisings from taking place. With Iraq reverting to its ethnic and sectarian roots, the Sunni Arabs viewing the struggle had significant reasons to fear for their future. Sunni Arabs had dominated Iraq since the beginning of the state and were in no hurry to test the mercy or toleration level of Iraq's Shi'ites should they seize power. While the Kurds sought autonomy or perhaps independence, the Shi'ites could be expected to have larger aims, including replacing the Sunnis as the political leadership of Iraq. There was also the possibility of a reign of terror unleashed by a Shi'ite government in response to the grinding oppression that Saddam had previously imposed on the Shi'ites, especially their religious leadership throughout his years in power. Militants among Iraq's Shi'ite clerics seemed likely candidates for national leadership since many secular Shi'ites had been compromised by collaboration with the regime.

Iraq's Sunni officers and military units overwhelmingly rallied to Saddam in response to this perceived sectarian threat, and the reaction against the twin uprisings was swift and brutal. In a clear representation of the sectarian viewpoint defining these events as well as Saddam's ongoing brutality, some Iraqi tanks in the south were painted with the slogan "No Shi'ites after today."[215] The level of carnage directed against defenseless people in the south was horrifying. Saddam's regime correspondingly survived because of Iraqi sectarian distrust and hatreds that never seemed to be far from the surface. The failure of the U.S. strategy to eliminate the regime led some analysts and politicians to conclude that Saddam had successfully deterred the United States from seizing Baghdad and thereby "finishing the job." This criticism evaporated after the 2003 U.S.-led invasion of Iraq and the subsequent problems empowering a new Iraqi government and suppressing a virulent insurgency.

After the 1991 war, various U.S. military leaders, including General Powell, stated that the United States was not contemplating the use of nuclear weapons in Iraq. Nevertheless, in Powell's account, the United States ruled out the use of nuclear weapons for operational and tactical reasons, not for strategic or moral reasons. He also described only one scenario under which nuclear weapons were ruled out involving the use of nuclear weapons against Iraqi armored forces on the battlefield, a role for which they clearly proved unnecessary.[216] Baghdad was probably oblivious to this nuance, but the Iraqis would probably not have cared if they did notice. The Iraqis, who were immersed in their own understanding of the use of propaganda, would never believe that Powell's statements were anything but the same type of self-serving propaganda that they would use if the situation was reversed.

CONCLUSION: LESSONS FOR OTHER CONFLICTS

The case studies of the 1973 Arab-Israeli War and the 1991 Gulf War provide valuable examples of the processes of escalation and intrawar deterrence that can occur in a regional conflict environment. It is important to understand how events unfolded during these conflicts to consider ways in which intrawar deterrent strategies might go well or poorly in future conflicts. It is also especially important to realize that any case study is limited in value by the special circumstances under which it occurred. Actions that occurred during these wars are important because they display a range of problems that can develop under similar circumstances or conflicts. Sweeping generalizations cannot be drawn from case studies such as these, although ways to think about future conflicts may be informed by these studies. It might also be noted that these cases must be understood in all their depth and nuance. Any effort to draw simple conclusions from a shallow understanding of these wars or to apply their lessons mechanistically is likely to lead to some flawed conclusions and results. Analogies have been consistently overused in the formation of U.S. policy often to the deep regret of the policymakers.[217]

In both of the conflicts under examination, the combatants did not use WMD, but in neither conflict was this restraint an inevitable result and some luck was involved in the outcomes. If the various journalistic and academic accounts can be believed, Israel may have come close to using nuclear weapons, but pulled back from this option because of the solid judgment of most of the Israeli top leadership and also because of the vast improvement of Israel's battlefield situation after October 14. Conversely, Saddam Hussein may have shown restraint because he had faith in his strategy to achieve his strategic objectives by conventional means. Saddam was deterred by U.S. threats and the probable belief that the United States was likely to follow through on those threats, but a more desperate leader may have responded in a different way. Thus, Saddam was prevented from using CBW by coalition threats but also by his own confidence in Iraq's conventional capabilities and a belief that the United States could not accept the type of prolonged ground war that he saw as required to oust the Iraqi regime. Saddam thus feared that the use of chemical or biological weapons would become a way to escalate the conflict from a level where he could remain in power to a new level where he could not.

In the future, it is at least possible that the United States will find itself in armed conflict against weaker nations that nevertheless possess WMD, perhaps including nuclear weapons. It is also possible that regional states using WMD will wage war against each other (for example, in South Asia). Some such conflicts may have a greater bearing on U.S. interests than others, but any nuclear exchange anywhere is of concern to global security. The use of biological weapons in combat would present its own special kind of nightmare should such actions serve as an example for other countries, and perhaps open a new and more hideous chapter in the history of warfare. Under these circumstances, wars involving vital U.S. interests (such as the 1991 Gulf War) may include an effort to engage in intrawar deterrence, but the confidence in this approach will have to be limited by the knowledge that escalation may become uncontrollable.

Iran is not a case study in this work, and Iran has not participated in a war similar to those discussed above. During the Iran-Iraq war, Tehran could not have responded to Iraqi actions with nuclear weapons once these became available. At that time, the only unconventional weapons that Iran possessed were limited stocks of chemical weapons which it employed in response to Iraqi chemical attacks. Throughout the war, Iran consistently lagged behind Iraq in CW capabilities. The experience of being on the losing end of the escalation ladder in its dealings with Iraq is nevertheless something that Tehran is unlikely to ignore in its future considerations about its national security. Moreover, the Tehran regime must have been shocked to the core by the post-1991 activities of the UN inspection teams in Iraq. Layer after layer of nuclear, biological, chemical, and missile infrastructure were rooted out by the UN, with the world, including Tehran, as spectators. Iran might have been particularly stunned by the knowledge that many of the fruits of these programs would have been directed at them had the Iraqi confrontation with the United States never occurred.

Future Iranian behavior in a time of crisis is difficult to predict, but aspects of both case studies may be worth considering in an Iranian context. Iran is neither a political democracy like Israel, nor is it a neo-Stalinist dictatorship under the iron rule of one individual such as existed in Saddam's Iraq. Still, some of the pathologies of Hussein's regime, including vastly distorted intelligence and insulated decision-making, could at least find a faint echo in Iranian crisis decision-making. The Iranians also have the potential to engage in a serious debate among decision-makers, but it is uncertain that they would do so in a way that headed off a catastrophic decision. It is also unclear if the factionalization of the Iranian political system would help or hurt such a debate. A serious debate on options such as apparently occurred in Israel in 1973 would be a challenging goal for Iran. Moreover, in a conflict with the United States or Israel, the Iranian leadership may feel the pressure to make almost instant decisions on WMD use.

The North Korean cultural and historical inputs to decision-making are profoundly different from those of Iraq but the political, intelligence-gathering, and governmental structure supporting decision-making appear to be quite similar to Iraq. Kim Jong-il sits atop the only contemporary regime that rivaled Saddam Hussein's government for the magnitude of the cult of personality. The ability of the North Korean system to generate rational decisions in an international crisis has most of the flaws of Saddam's decision-making process, and may perhaps involve greater degrees of recklessness than found in the Iraqi system. It is possible that Saddam was both smarter and less deluded than Kim Jong-il, but it is also likely that the North Koreans have deadlier weapons and more military options than Iraq did in 1991. These include nuclear weapons, chemical weapons, possible biological weapons, and conventional strikes across the border with South Korea. Saddam's strategy in the struggle with the United States was flawed but based on a serious effort to deter a U.S. attack and then at least save the regime if the United States was undeterable. It is unclear that the North Koreans can manage a strategy that is even this coherent. This situation suggests that any future confrontation with North Korea will be much more difficult to manage than the Iraqis were in 1991. A surgical strike against North Korea may meet a wholly unexpected response that an American strategist would never consider realistic for Pyongyang.

The following recommendations are offered regarding the above issues. Some of these recommendations can be seen to draw more heavily from one of the case studies than the other. This is to be expected since the two case studies were chosen not only for their relevance, but for the profound differences between them. All of these recommendations are meant to apply most directly to limited war situations where intrawar deterrence is an issue. Additionally, these recommendations do not always assume that the United States is a party to the conflict. In many cases, the United States will need to consider the actions that various states may take against each other (say India and Pakistan) in order to understand if the conflict is likely to escalate or recede. In some cases, U.S. diplomacy may be informed by a solid understanding of the potential variation in the escalatory process as U.S. leaders work with the UN to try to establish cease-fires prior to WMD use.

1. U.S. policymakers need to remain cognizant that limited military attacks may not appear limited to those nations under attack. In 1973, some intelligent and experienced Israeli leaders believed that they faced an existential threat, although most Israeli and other historians with the benefit of time and study no longer support this view. In contemporary times, large-scale attacks can start to look like an effort at regime change.

The temptation for foreign nations to respond to perceived regime changing attacks with every resource available will be serious. While escalation was avoided in the 1973 and 1991 wars, reasons for this restraint might not always be present. The Israeli government, drawing upon its democratic principles, engaged in an open and rigorous debate on escalation issues in which a moderate majority swayed the Prime Minister into a better understanding of the military situation and helped to neutralize the unrelenting pessimism attributed to Defense Minister Dayan. As noted, countries such as Iran also have a tradition of governmental debate, but it does not rise to the Israeli standard. A limited U.S. attack against Iran or North Korea could be viewed as the beginnings of an existential challenge to these regimes, although this interpretation may be more likely with Pyongyang than Tehran since that regime is by far the most insulated and paranoid of the two. Nevertheless, even an Israeli attack against Iran could be viewed as the beginning of a U.S.-Israeli campaign to destroy the Islamic Republic, and it could provoke an overwhelming response.

2. U.S. policymakers need to remain aware that intrawar deterrence is an inherently fragile concept, and that no plan survives first contact with the enemy. One power engaged in the conflict may engage in "signaling," only to have that signal completely misunderstood by the other party or parties to the conflict. The problem with sending one set of signals to multiple audiences, including adversaries and allies, might also cause a tendency toward vagueness which may be understood in completely wrongheaded ways by adversaries. The United States military leadership should consequently not allow itself to become too comfortable with the concept of intrawar deterrence in the military planning process. It must be fully understood that it is an approach which may or may not work depending on a variety of factors that occur throughout a crisis and war. The temptation for adversary states to strike back with some or all of their weapons to arrest a declining conventional situation will always have to be considered.

3. Military planners must remain cognizant that issues regarding the "fog of war" remain continuously relevant throughout contemporary conflicts. In the 1973 war, the Israelis had to cope with both failed military doctrine and serious military setbacks at the same time. They had to adapt to alarmingly new conditions as they were experiencing massive military losses. Dayan's reported statement that he was wrong about everything may or may not have actually been spoken, but it reflected a serious mindset among Israeli decision-makers that could become a problem for future wartime leaders. What do you do when you find out that you were wrong about everything? What decisions will you make when you fall into a valley of unrelenting pessimism?

4. All those involved with U.S. national strategy must remain aware that perceptions will remain the core of most intrawar deterrence situations. Different individuals will clearly see "reality" with different levels of accuracy. Even bright and experienced people may take some time before they achieve a reasonable level of situational awareness during a crisis. When looking at the military situation on the Suez Canal, Ariel Sharon immediately understood that the Egyptians were not capable of a ground war that moved beyond the protection of their air defense systems over the western Sinai. Less gifted commanders such as Goren did not understand that and made severe operational mistakes, including a hasty and inadequately resourced counterattack as a result. Moreover, less gifted generals report their activities and progress upwards in ways that distort the perceptions of their superiors. The United States must recognize, as the Egyptians did not in 1973, that actions that are self-evident to one party in a conflict may or may not be self-evident to other parties in that conflict. U.S. leaders also need to be aware that emerging nuclear weapons states may be developing their nuclear weapons doctrine "on the fly" during future conflicts.

5. Military planners may need to be particularly aware that deterrence during wartime situations is an ongoing and evolving process that may need to be adjusted as the war continues and the military circumstances of various conventional forces change. States that fully plan and expect to fight wars on a conventional level can find themselves scrambling for options once pre-war assumptions fail. In some cases, WMD options could be used in ways that have not been addressed by doctrine or strategic analysis and the consequences of which have not always been scrutinized with the depth that they deserve. The rapidly changing natures of the WMD threats in both case studies underscore how rapidly new situations develop in wartime situations, and how threats of WMD use go from minimal to serious in a short period of time. The Israelis may not have thought very seriously about the circumstances under which they would use nuclear weapons until they were in the middle of a new and particularly challenging war.

6. U.S. leaders involved in or viewing a conflict must understand the communications can be awkward and blundering in situations of intrawar deterrence. Signaling in general is not a highly nuanced or inherently accurate form of communications. Some communications will probably be misunderstood, downplayed, or assumed away if they are not continuously reinforced and emphasized. The U.S. decision to emphasize the dire consequences of Saddam Hussein's chemical weapons use may not have been the only major factor preventing an Iraqi use of chemical and biological weapons, but it certainly was a worthwhile exercise.

7. U.S. leadership cannot take the good judgment of foreign leaders for granted. In times of war, individuals with a lifetime of experience in security matters can still make fundamental mistakes about the nature of the conflict with which they have become involved. In the 1973 October War, the Israelis had an especially difficult challenge in showing the flexibility to overcome the tendency to fall back on the experience of the victorious 1967 war. They also had to discern Arab intensions within their own psychological framework which emphasized the searing memory of the most tragic episodes in the history of the Jewish people and their more recent experiences, including a great deal of inflammatory Arab propaganda. In Saddam Hussein's case, the problem was much more acute since he sat on top of a dictatorial system that indulged his delusions and shielded him from potentially unpleasant truth.

8. U.S. analysts looking at actual or potential military conflict must guard against the tendency to view wartime behavior in ways that minimize the linkage of wartime behavior and regime survival. This approach is, of course, difficult since regimes have different survival strategies. Some wartime adversaries will attempt to make decisions that are both in the interests of their country and in the interests of their own regime. Some leaderships such that of North Korea may see little value in the national survival of their country unless the regime survives as well. Wartime efforts to encourage a military coup may be useful, but the example of the 1991 Gulf War illustrates both the difficulty of such an action and the likelihood that unexpected variables may come into play. In general, it may be a bad idea to assume that the massive military defeat of most countries will lead to an automatic coup against the ruling regime in a limited war scenario.

9. U.S. planners must remain aware of the ways in which information is obtained, processed, and presented to foreign leaders. The intelligence services of foreign nations may not be providing a foreign leadership with accurate intelligence on which its leadership can base a reasonable cost-benefit decision on whether or not to use unconventional weapons. Foreign intelligence services may be giving a foreign leadership a completely wrongheaded view of how a conflict is evolving. In some cases, such as in 1991 Iraq, intelligence services can be unprofessional, corrupt, and incompetent. This certainly was the case with Saddam's intelligence service which catered to his prejudices and personality problems. While the logic of intrawar deterrence is transnational, effective communications requires an understanding of the audience or audiences that one is seeking to influence.

ENDNOTES

1. Egypt and Syria did the bulk of the fighting on the Arab side although meaningful contributions of military forces came from Iraq, Jordan, and Morocco. Some other Arab nations also contributed smaller numbers of forces. On limited Arab objectives, see Henry Kissinger, *Years of Upheaval*, Boston, MA: Little, Brown and Company, 1982, pp. 459-460.

2. Israeli Prime Minister Levi Eshkol once characterized Israeli psychology in Yiddish as *Shimshon der nebechdikker*. This phrase is very roughly translated by Israeli historian (and current Israeli Ambassador to the United States) Michael B. Oren as "Samson the nerd." Eshkol was referring to the tendency of many Israelis to swing between the belief that their country is militarily invincible and deeply vulnerable. Michael B. Oren, *Six Days of War: June 1967 and the Making of the Modern Middle East*, Oxford, United Kingdom (UK): Oxford University Press, 2000, p. 18.

3. Thomas C. Schelling, *Arms and Influence*, New Haven, CT, and London, UK: Yale University Press, 1966, p. 135.

4. See for example, Avner Yaniv, *Deterrence Without the Bomb*, Lexington, MA: Lexington Books, 1987.

5. Patrick M. Morgan, *Deterrence Now*, Cambridge, UK: Cambridge University Press, 2003, p. 58.

6. Zeev Maoz, *Defending the Holy Land: A Critical Analysis of Israel's Security & Foreign Policy*, Ann Arbor: University of Michigan Press, 2009, p.305-306.

7. *Ibid*., p. 306, Shlomo Aronson with Oded Brosh, *The Politics and Strategy of Nuclear Weapons in the Middle East: Opacity Theory, and Reality 1960-1991: An Israeli Perspective*, New York: State University of New York Press, 1992, chapter 9.

8. Maoz, p. 320.

9. For a comprehensive look at the Six Day War, see Oren, *Six Days of War*.

10. For more information, see W. Andrew Terrill, "Iraq's Role in the 1973 Arab-Israeli War," *Small Wars and Insurgencies*, Vol. 11, No. 3, Winter 2000, p. 5.

11. Nir Hefez and Gadi Bloom, *Ariel Sharon: A Life*, New York: Random House, 2006, p. 108.

12. In the mid-1990s a retired Egyptian general who served as a junior officer in 1967 told the author that his own children

made jokes to him about Egyptian military incompetence. These exchanges, however shocking within Arab culture, are by no means isolated instances since the defeat produced widespread masochistic black humor throughout the country.

13. Abraham Rabinovich, *The Yom Kippur War: The Epic Encounter that Transformed the Middle East*, New York: Schocken Books, 2004, p. 7.

14. Insight Team of the *Sunday Times*, *The Yom Kippur War*, New York: ibooks and Simon and Schuster, 2002, p. 95.

15. On July 25, 1970, five Soviet pilots were shot down by Israel aircraft. There were no Israeli losses, and only one Israeli Mirage fighter was damaged in this encounter. See Lon Nordeen and David Nicolle, *Phoenix Over the Nile: A History of Egyptian Air Power 1932-1994*, Washington, DC, and London, UK: Smithsonian Institution Press, 1996, p. 253.

16. Soviet instruction for Egyptian pilots was often arrogant and insulting, leading some resentful Egyptian officers to take a bit of satisfaction from the Soviet humiliation at Israeli hands. See *Ibid*, p. 253.

17. The valuable intelligence work done by one young intelligence officer, Lieutenant Benjamin Siman-Tov, was, for example, suppressed by the leadership of Israeli military intelligence. See Insight Team, p. 107.

18. Rabinovich, *The Yom Kippur War*, p. 56. Also see Yaacov Bar-Siman-Tov, *The Israeli-Egyptian War of Attrition 1969-1970*, New York: Columbia University Press, 1980, chapter 5.

19. Oren, p. 176.

20. Zeev Schiff, *October Earthquake: Yom Kippur 1973*, New Brunswick, NJ: Transaction Books, 1974, p. 25.

21. George W. Gawrych, *The Albatross of Decisive Victory: War and Policy Between Egypt and Israeli in the 1967 and 1973 Arab-Israeli Wars*, Westport, CT: Greenwood Press, 2000, pp. 163-164.

22. Rabinovich, *The Yom Kippur War,* p. 144; Kenneth M. Pollack, *Arabs at War: Military Effectiveness, 1948-1991,* Lincoln and London, UK: University of Nebraska Press, 2002, pp. 481-482.

23. Sadat and other Egyptian leaders complained bitterly throughout the war and afterwards that the United States was providing unlimited support to Israel, while the Soviet Union was parsimonious with its own level of support to Egypt.

24. Aronson, p. 144.

25. In contrast to Egypt, serious Israeli-Syrian negotiations did not begin until the early 1990s. These negotiations failed. See Itamar Rabinovich, *The Brink of Peace: The Israeli-Syrian Negotiations,* Princeton, NJ: Princeton University Press, 1998.

26. Maoz, p. 152.

27. *Ibid.*

28. Pollack, pp. 481-482.

29. Nigel Ashton, *King Hussein of Jordan: A Political Life,* New Haven, CT, and London, UK: Yale University Press, 2008, p. 185.

30. Martin van Creveld, *Nuclear Proliferation and the Future of Conflict,* New York: The Free Press, 1993, pp. 108-110.

31. Avner Cohen, *Israel and the Bomb,* New York: Columbia University Press, 1998, pp. 273-276.

32. For a wide assortment of declassified U.S. Government documents on this issue, see Avner Cohen, ed., "Israel Crosses the Threshold," National Security Archive, April 28, 2006, *www.gwu.edu/-nsarchiv.*

33. Some limited U.S. inspections of Dimona did occur but were so limited that it is difficult to take them very seriously.

34. Lieutenant General Saad al Shazly, *The Crossing of the Suez,* San Francisco, CA: American Mideast Research, 2003, pp. 78-81.

35. *Ibid.*, pp. 77-78.

36. *Ibid.*, p. 116.

37. *Ibid.* Note that at this time General (later President) Hosni Mubarak served as Air Force Chief of Staff under Baghdadi and was thus the second highest ranking officer in the Air Force.

38. Around 15,000 Soviet advisors were expelled. Around 200 technicians remained, in addition to a limited number of instructors at the Nasser Higher Academy. T. N. Dupuy, *Elusive Victory: The Arab-Israeli Wars, 1947-1974*, Dubuque, IO: Kendall/ Hunt Publishing Company, 1992, p. 390.

39. Mohamed Abdel Ghani El-Gamasy, *The October War: Memoirs of Field Marshal El-Gamasy*, Cairo, Egypt: American University of Cairo Press, 1993, p. 144.

40. Mohamed Heikal, *The Sphinx and the Commissar: The Rise and Fall of Soviet Influence in the Middle East*, New York: Harper and Row, 1978, chapter 15.

41. El-Gamasy, p. 147.

42. Steven J. Zaloga, *Scud Ballistic Missile and Launch Systems, 1955-2005*, New York: Osprey Books, 2006, p. 43.

43. *Ibid.*

44. This monograph will use the standard U.S. Army/U.S. Air Force definitions of missiles as systems which have internal guidance systems and rockets as systems that do not.

45. These toxins may be drawn from living plants or animals, or they may be synthetically reproduced.

46. Stockholm International Peace Research Institute, *The Problem of Chemical and Biological Warfare, Volume II: CB Weapons Today*, New York: Humanities Press, 1973, p. 241.

47. *Ibid.*, p. 241.

48. Jerry Asher with Eric Hammel, *Duel for the Golan*, New York: William Morrow and Company, Inc, 1987, p. 61.

49. Rabinovich, *The Yom Kippur War*, p. 98. Mendler was killed later during the war in the Sinai when an artillery shell hit his command half track.

50. Evron, p. 73.

51. Rabinovich, *The Yom Kippur War*, p. 142.

52. Schiff, p. 40; and Avraham (Bren) Adan, *On the Banks of the Suez: An Israeli General's Personal Account of the Yom Kippur War*, Novato, CA: Presidio Press, 1991, p. 162.

53. Insight Team, p. 139.

54. Rabinovich, *The Yom Kippur War*, p. 156.

55. *Ibid*, p. 261.

56. *Ibid*, p. 206.

57. Schiff, p. 59.

58. Rabinovich, *The Yom Kippur War*, p. 115.

59. Walter J. Boyne, *The Two O' Clock War: The 1973 Yom Kippur Conflict and the Airlift that Saved Israel*, New York: St. Martin's Press, 2002, p. 31.

60. Rabinovich, *The Yom Kippur War*, p. 171.

61. Golda Meir, *My Life*, New York: Dell, 1975, p. 413.

62. Schiff, p. 92.

63. Hanoch Bartov, *Dado: 48 Years and 20 Days: The Full Story of the Yom Kippur War and of the Man Who Led Israel's Army*, Tel Aviv, Israel: Ma'ariv Book Guild, 1981, pp. 328-329. Note that Bartov was working with Elazar on his memoirs when Elazar died of

a fatal heart attack in 1976. This book is based on the completed preliminary draft Elazar produced and remains the most accurate available representation of Elazar's version of his life and events.

64. *Ibid.* p. 329.

65. Rabinovich, *The Yom Kippur War,* p. 219.

66. Lieutenant Colonel David Eshel, Israeli Defense Forces, Retired, "Counterattack in the Sinai, 8 October 1973," *Military Review*, November 1993, pp. 54-66.

67. Schiff, p. 113.

68. Rabinovich, *The Yom Kippur War,* p. 225.

69. General Gonen's reputation and standing within Israel never recovered from the 1973 war. As a result, he sought a new start working in the Central African Republic. See van Creveld, *Moshe Dayan*, p. 178.

70. Asher (with Hammel), p. 178.

71. *Ibid.* p. 140.

72. Insight Team, p. 193.

73. Kissinger, p. 492.

74. Elinor Burkett, *Golda*, New York: HarperCollins Publishers, 2008, p. 323.

75. Insight Team, p. 190.

76. Ashton, Chapter 9.

77. Chaim Herzog, *The War of Atonement*, Boston, MA: Little, Brown and Company, 1975, pp. 139-140.

78. Rabinovich, *The Yom Kippur War,* p. 90.

79. Burkett, p. 339.

80. "How Israel Got the Bomb," *TIME*, April 12, 1976, p. 39.

81. Avner Cohen, "The Last Nuclear Moment," *The New York Times*, October 6, 2003, internet.

82. Aronson (with Brosh).

83. Herzog, p. 116.

84. Blum, p. 228.

85. Aronson, pp. 146-147.

86. *Ibid.*, p. 147.

87. *Ibid.*

88. "Yuval Ne'eman: 'During the Yom Kipper War, we deployed Jericho missiles, and left it to the Egyptians to guess what they were carrying'," *Ha'aretz*, May 23, 1997, internet.

89. See Avner Cohen, "Israel and Chemical/Biological Weapons: History, Deterrence and Arms Control," *The Nonproliferation Review*, Fall-Winter 2001, pp. 27-53. According to Dr. Cohen, "To this day, Israeli's past and present activities in the CBW [Chemical and Biological Warfare] field are treated as a national taboo—similar to, and perhaps even stronger than the nuclear taboo." P. 34.

90. Van Creveld, *Moshe Dayan*, p. 167.

91. Van Creveld, *Nuclear Proliferation and the Future of Conflict*, p. 102.

92. *Ibid*, p. 102; and C. Wakebridge, "The Syrian Side of the Road," *Military Review*, February 1976, p. 29.

93. Yair Evron, *Israel's Nuclear Dilemma*, Ithaca, NY: Cornell University Press, 1994, pp. 71-72.

94. Moshe Dayan, *Story of My Life,* New York: Warner Books, 1976, p. 598.

95. Rabinovich, *The Yom Kippur War,* p. 231.

96. Burkett, p. 324.

97. Dayan interview, A&E television biography series, *Moshe Dayan, A Warrior's Story,* 1996.

98. Seymour M. Hersh, *The Samson Option, Israel's Nuclear Arsenal and American Foreign Policy,* New York: Random House, 1991, chapter 17.

99. Depuy, p. 569.

100. While Defense Secretary Schlesinger initially wanted to provide more modest levels of support to Israel, Kissinger basically was allowed to decide the policy. He may have used rumors of bureaucratic conflict to his own advantage in negotiating with the Israelis. See Robert Dallek, *Nixon and Kissinger Partners in Power,* New York: HarperCollins, 2007, p. 523; and Alistair Horne, *Kissinger 1973, The Crucial Year,* New York: Simon & Schuster, 2009, p. 244.

101. Schiff, p. 168.

102. Kissinger, p. 462.

103. Cited in Alistair Horne, *Kissinger 1973,* New York: Simon & Schuster, 2009, p. 233.

104. Dupuy, p. 465.

105. See comments by Dayan as cited by Insight Team, p. 214.

106. Dupuy, p. 467.

107. Victor Israelyan, *Inside the Kremlin During the Yom Kippur War,* University Park: Pennsylvania State University Press, 1995, p. 68.

108. *Ibid*, p. 17.

109. Mohamed Heikal, *The Road to Ramadan*, New York: Ballantine Books, 1975, p. 220.

110. Dupuy, p. 465.

111. Israelyan, p. 67.

112. Insight Team, p. 315.

113. Dupuy, p. 533.

114. Shazly, pp. 246-248.

115. See Kissinger, pp. 459-460.

116. Dupuy, p. 487.

117. Burkett, p. 332.

118. Rabinovich, *The Yom Kippur War*, p. 415.

119. Insight Team, p. 411-412.

120. Anwar el Sadat, *In Search of Identity: An Autobiography*, New York: Harper & Row, 1977, p. 265.

121. "How Israel got the Bomb," p. 39.

122. William B. Quandt, "Soviet Policy in October Middle East War—II," *International Affairs*, October 1977, p. 596.

123. *Ibid*, p. 596.

124. *Ibid.*, pp. 596-597.

125. Cecil Brownlow, "Soviets Poise Three Front Global Drive," *Aviation Week and Space Technology*, November 5, 1973, p. 12.

126. Wendell Stevenson, *The Weight of a Mustard Seed: The Intimate Story of an Iraqi General and His Family During Thirty Years of Tyranny*, New York: Harper Collins, 2009, p. 77. Such a comment is, of course, true. The culture clash between brash, straightforward Americans and much more subtle, indirect, and protocol conscious Arabs is often maddening for both parties. On one occasion, an Iraqi general stated that he did not wish to discuss serious matters with me because this was our first meeting, and such discussions were not appropriate until I made a return trip to Baghdad. Other senior Iraqi officers have been much more straightforward, perhaps in response to the importance of the military issues that need to be discussed or the differences between Iraqi culture overall and the subculture of the military.

127. *Ibid*, p. 65.

128. *Ibid*, p. 78.

129. For an especially good analysis of Saddam Hussein's personality, see Jerrold M. Post, M.D., and Amatzia Baram, Ph.D., *"Saddam is Iraq: Iraq is Saddam,"* Maxwell Air Force Base, AL: USAF Counterproliferation Center: Future Warfare Series No. 17, November 2002, especially pp. 1-9. Note that Professor Post is a former CIA psychiatrist who has devoted his career to the study of political psychiatry.

130. Kanan Makiya, *Republic of Fear, The Politics of Modern Iraq*, Berkeley: University of California Press, 1998, p. 119.

131. Ashton, p. 266.

132. Alan Munro, *Arab Storm, Politics and Diplomacy Behind the Gulf War*, London, UK, and New York: I. B. Tauris, 2006, p. 7. Note that Munro was the United Kingdom's Ambassador to Saudi Arabia during the 1991 war.

133. For an example, see Stevenson, p. 53.

134. Richard N. Haass, *War of Necessity, War of Choice*, New York: Simon & Schuster, 2009, p. 56.

135. Munro, p. 29.

136. "The Glaspie Transcript: Saddam meets the U.S. Ambassador" (July 25, 1990), Michah L. Sifry and Christopher Cerf, eds., *The Gulf War Reader, History, Documents, Opinions,* New York: Random House, 1991, pp. 122-133; and Glenn Kessler, "Ex-envoy Details Hussein Meeting," *The Washington Post,* April 3, 2003, internet.

137. Norman Cigar, "Iraq's Strategic Mindset and the Gulf War: Blueprint for Defeat," Journal *of Strategic Studies,* March 1992, p. 4.

138. Cooley, p. 188.

139. See Colin Powell, *My American Journey,* New York: Ballentine Books, 1995, p. 463; Haass, p. 63.

140. James A. Baker III, *The Politics of Diplomacy, Revolution, War & Peace,* New York: G. P. Putnam's Sons, 1995, p. 276.

141. Freeman and Karsh, p. 218.

142. Haass, p. 77.

143. *Ibid.*, p. 79.

144. Kevin M. Woods, *The Mother of all Battles, Saddam Hussein's Strategic Plan for the Persian Gulf War,* Annapolis, MD: Naval Institute Press, 2008, p. 129. Please note that this study is based almost entirely on captured Iraqi documents and interviews with former Iraqi leaders. Many of the documents related to the 1991 War were seized in the aftermath of the 2003 Iraq War.

145. Lawrence Freedman and Efraim Karsh, *The Gulf Conflict, 1990-1991,* Princeton, NJ: Princeton University Press, 1993, p. 377.

146. Richard P. Hallion, *Storm over Iraq: Air Power and the Gulf War,* Washington, DC, and London, UK: Smithsonian Institution Press, 1992, p. 172.

147. Woods, *The Mother of all Battles,* p. 197.

148. *Ibid.*, p. 137.

149. See U.S. Embassy Cable to Secretary of State (071310Z March 1984) declassified and placed on the National Security Archives Website.

150. Richard L. Russell, "Iraq's Chemical Weapons Legacy: What Others Might Learn from Saddam," *Middle East Journal*, Spring 2005, p. 197.

151. Joost R. Hiltermann, *A Poisonous Affair: America, Iraq and the Gassing of Halabja*, Cambridge, UK, and New York: Cambridge University Press, 2007, p. 12.

152. Warren W. Lenhart and Todd Masse, *Persian Gulf War: Iraqi Ballistic Missiles Systems: CRS Report for Congress*, Washington DC: Congressional Research Service, 1991, p. 1.

153. W. Seth Carus and Joseph S. Bermudez, Jr., "Iraq's al Husayn Missile System, Part 2," *Jane's Soviet Intelligence Review*, June 1990, p. 205.

154. As cited in Timothy V. McCarthy and Jonathan B. Tucker, "Saddam's Toxic Arsenal: Chemical and Biological Weapons in the Gulf Wars," Peter R. Levoy, Scott D. Sagan, and James J. Wirtz, *Planning the Unthinkable*, Ithaca, NY, and London, UK: Cornell University Press, 2000, p. 58.

155. Powell, p. 468.

156. See, for example, Seth Carus, *The Genie Unleashed: Iraq's Chemical and Biological Weapons Production*, Washington DC: Washington Institute Policy Papers, No. 14, 1989.

157. For an array of declassified U.S. Government documents on the development of Iraqi WMD and their delivery systems, see Jeffrey Richelson, ed., "Iraq and Weapons of Mass Destruction," *National Security Archive Electronic Briefing Book No. 80*, updated February 11, 2004, *www.gwu.edu/-nsarchiv*.

158. See UN Special Commission (UNSCOM), "Status of Verification of Iraq's Biological Warfare Programme," January

25, 1999, as cited on Federation of American Scientists website, *www.fas.org*; and Jean E. Krasno and James S. Sutterlin, *The United Nations and Iraq*, Westport, CT: Praeger, 2003, p. 61.

159. Richard Butler, *The Greatest Threat: Iraq, Weapons of Mass Destruction, and the Crisis of Global Security*, New York: Public Affairs Publishers, 2000, p. 82.

160. *Ibid.*, p.123.

161. Khidhir Hamza, *Saddam's Bombmaker*, New York: Scribner, 2000, p. 237.

162. The author looked at this in some depth in W. Andrew Terrill, "Chemical Weapons in the Gulf War," *Strategic Review*, Spring 1986, pp. 51-58.

163. John Tagiabuf, "Iraqi Weapons Had Chemical Warheads," *New York Times*, November 12, 1991, internet.

164. Bob Drogin, *Curveball: Spies, Lies, and the Con Man Who Caused a War*, New York: Random House, 2007, pp. 47-48.

165. After the 2003 war, some improperly disposed remnants from the old WMD programs were found, but nothing that could be considered usable weapons.

166. Baram, p. 30.

167. Hamza, p. 129.

168. Moaz, p. 317.

169. *Ibid.*

170 . As cited in Efraim Inbar, *Rabin and Israel's National Security*, Washington DC: The Woodrow Wilson Center Press, 1999, p. 127.

171. *Ibid.*

172. Vanunu was released in 2004 after serving his sentence but remains subject to a number of restrictions on his speech and travel. Maoz, p. 305.

173. As cited in Kevin M. Woods, *The Mother of all Battles, Saddam Hussein's Strategic Plan for the Persian Gulf War*, Annapolis, MD: Naval Institute Press, 2008, p. 18.

174. Woods, *The Mother of all Battles*, p. 18.

175. *Ibid.*, p. 26.

176. Baghdad Iraqi News Agency, August 18, 1990, as reported in Foreign Broadcast Information Service (FBIS), "INA Reports Aziz Remarks on Chemical Weapons," *Daily Report Near East and South Asia*, August 20, 1990, p. 17.

177. Paris Radio Monte Carlo as cited in *Chemical Weapons Convention Bulletin*, Issue 11, March 11, 1991, p. 10.

178. China was the only major power consistently willing to sell combat aircraft to the Iranians during this war. See Anthony H. Cordesman, *Iran and Iraq: The Threat from the Northern Gulf*, Boulder, CO: Westview Press, 1994, pp. 54-55.

179. Woods, *The Mother of all Battles*, p. 78.

180. *Ibid.*, p. 220.

181. Laura King, "Iraq Vows Missiles Will Rain on Neighbor that Permits Invasion," *Washington Times*, November 26, 1990, p. A-8.

182. David Lamb, "Iraq Launches 3rd Missile—but not at Allied Forces," *Los Angeles Times*, December 29, 1990, p. 15.

183. Amman Jordan Television Network, January 9, 1991, in FBIS, "Tariq Aziz Geneva Press Conference," *Daily Report: Near East & South Asia*, January 10, 1991, p. 44.

184. Lamb, p. 15.

185. Edward Cody, "Saddam Claims to Have New Missile that Can Hit Israel," *Washington Post*, October 10, 1990, p. A-19.

186. R. Jeffery Smith, "US Warns of Retaliation if Iraq uses Poison Gas," *Washington Post*, August 9, 1990, A-33.

187. "News Chronology," *Chemical Weapons Convention Bulletin*, Issue 11, March 1991, pp. 7-8.

188. *Ibid.*, p. 8.

189. Norman Cigar, "Chemical Weapons and the Gulf War: The Dog that did not Bark," *Studies in Conflict and Terrorism*, Spring 1992, p. 149.

190. Bush letter to Saddam as cited in Michah L. Sifry and Christopher Cerf, eds., *The Gulf War Reader, History, Documents, Opinions*, New York: Random House, 1991, p. 179.

191. Baker, p. 359.

192. *Ibid.*, p. 359.

193. William Touhy, "Quayle Refuses to Rule Out U.S. Nuclear Attack," *Los Angeles Times*, February 2, 1991, p. A-10.

194. Interview with Secretary of Defense Cheney on "Evans and Novak," as cited in William M Arkin, "Calculated Ambiguity" Nuclear Weapons and the Gulf War," *Washington Quarterly*, Autumn 1996, pp. 6-7.

195. "Excerpts from President Bush's News Conference," *Washington Post*, February 6, 1991, p. A-21.

196. Warren Strobel, "Rules Change if Iraq uses Gas weapons," *Washington Times*, February 8, 1991, p. A-1.

197. Saddam Hussein's January 28, 1991, interview, CNN Video, "War in the Gulf," Part 4, 1991.

198. Saudi Arabian Television, February 3, 1991, "Further Reportage on Allied Actions in the Gulf War Discussion of Iraqi Gas Masks," as cited in FBIS, *Daily Report on the Middle East & South Asia*, February 4, 1991, p. A-21.

199. Cigar, "Chemical Weapons and the Gulf War," p. 155. For an example of one of the ads, see *Washington Times*, October 23, 1990 p. A-9.

200. Woods, *The Mother of All Battles*, p. 126.

201. For examples of these instructions from newspapers such as al-Qadisiyah and al-Iraq in addition to various radio broadcasts, see FBIS, *Daily Report Near East and South Asia*, December 24, 1990, pp. 26-29.

202. Woods, *The Mother of all Battles*, p. 182.

203. Efraim Halevy, *Man in the Shadows: Inside the Middle East Crisis with a Man Who Led the Mossad*, New York: Saint Martin's Press, 2006, pp. 31-32; Amatzia Baram, "An Analysis of Iraqi WMD Strategy, *The Nonproliferation Review*, Summer 2001, p. 29; Woods, *The Mother of all Battles*, p. 149.

204. Rick Atkinson, *Crusade: The Untold Story of the Persian Gulf War*, Boston and New York: Houghton Mifflin Company, 1993, pp. 82-83.

205. Halevy, p. 31.

206. *Ibid.*, p. 32.

207. Atkinson, p. 82.

208. See Yitzhak Shamir, *Summing up: An Autobiography*, Boston, MA: Little, Brown and Company, 1994, p. 217.

209. Woods, *The Mother of all Battles*, p. 189.

210. Cigar, "Chemical Weapons and the Gulf War," pp. 145-155.

211. On the battle itself, see Atkinson, pp. 201-213.

212. Freedman and Karsh, p. 377.

213. *Ibid.*, p. 380.

214. George H. W. Bush and Brent Scowcroft, *A World Transformed*, New York: Alfred A. Knopf, 1998, p 472.

215. Kanan Makiya, *Cruelty and Silence: War, Tyranny, Uprising, and the Arab World*, New York: W. W. Norton & Company, 1991, p. 97.

216. Powell, p. 486.

217. For an especially thoughtful examination of this problem, see Jeffrey Record, *Making War, Thinking History: Munich, Vietnam, and Presidential Uses of Force from Korea to Kosovo*, Annapolis, MD: Naval Institute Press, 2002.